Selection of Representative TPH Fractions Based on Fate and Transport Considerations

Total Petroleum Hydrocarbon Criteria Working Group Series

Volume 3

Selection of Representative TPH Fractions Based on Fate and Transport Considerations

JULY 1997

PREPARED FOR:
Association of American Railroads, United States Air Force, and the Total Petroleum Hydrocarbon Criteria Working Group

BY:
John B. Gustafson, Ph.D.
Shell Development Company

Joan Griffith Tell, Ph.D.
Exxon Biomedical Sciences, Inc.

Doug Orem
EA Engineering, Inc.

Amherst Scientific Publishers
150 Fearing Street
Amherst, Massachusetts 01002

ISBN 1-884-940-12-9

A portion of the proceeds from the sale of this book will be donated to the Plant-a-Tree Program, a reforestation program managed by the U.S. Forest Service.

Printed in the United States of America

CONTENTS

PREFACE

This document is third in a series from the Total Petroleum Hydrocarbon Criteria Working Group (TPHCWG). The Working Group was formed in 1993 based on the observation that widely different clean-up requirements were being used by states at sites that were contaminated with hydrocarbon materials such as fuels, lubricating oils, and crude oils. These requirements were usually in the form of concentrations of total petroleum hydrocarbon, otherwise known as TPH, and ranged from 10 to over 10,000 milligrams of hydrocarbons per kilogram of soil. Members of the group jointly recognized that the numerical standard was not based on a scientific assessment of human health risk and they established the following goal for their effort:

> *To develop scientifically defensible information for establishing soil cleanup levels that are protective of human health at hydrocarbon contaminated sites.*

The Working Group is guided by a steering committee consisting of representatives from industry, government, and academia. Some of the active participants among the more than 400 involved, include the Gas Research Institute, the Petroleum Environmental Research Forum, several major petroleum companies including Chevron, Exxon, and Shell, the American Petroleum Institute, the Association of American Railroads, several state governments (i.e., Washington, Texas, Colorado, Hawaii, Louisiana, New Mexico), the U.S. Environmental Protection Agency, the Department of Defense, and many consulting firms such as EA Engineering, Science and Technology.

An overlying theme to this document is the importance of *exposure* potential when defining human health risk. The fate and transport of a chemical or mixture defines the exposure route and, in conjunction with receptor properties, concentrations at receptors. If fate and transport is not considered, unrealistic human health risks could be calculated, resulting in misinformed decisions about site clean-up, regulatory guidance, etc.

This document summarizes the methods used to delineate TPH into equivalent carbon number fractions based on fate and transport considerations. The input into the fraction method included composition data on many common fuels and petroleum products. This information is provided in detail in Volume 2 of the Working Group reports. Once the fractions were defined, fraction-specific values of relevant physical-chemical properties were calculated based on correlations to boiling point. Companion volumes include Volume 1 which provides an overview of the complexities of petroleum hydrocarbon characterization and risk assessment and a discussion on the analytical methods available. In addition to descriptions about general analytical methods we have also provided a summary of a proposed GC-based analytical method developed by the Working Group that reports hydrocarbon results in equivalent carbon number groups or fractions.

To complete the risk-based approach, the Working Group has also selected toxicity criteria (e.g., Reference Doses) for each of the defined fate and transport fractions. The evaluation of the toxicology research database and rationale behind the toxicity criteria selected is described in detail in Volume 4, "Development of Fraction-Specific Toxicity Criteria for Total Petroleum Hydrocarbons (TPH)" (in preparation). The analytical method, fate and transport considerations and toxicity criteria are the technical elements which fit into a risk-based framework for determining human health based criteria at petroleum hydrocarbon contaminated sites. The group selected the American Society for Testing and Materials (ASTM) Risk Based Corrective Action - RBCA framework as an example of how these elements can be used to calculate risk-based screening levels driven by noncancer human health risk for petroleum contaminated sites. We hope you find this document to be useful in your efforts to evaluate and determine acceptable risk-based criteria at petroleum sites.

ACKNOWLEDGMENTS

The publication of this volume of the Total Petroleum Hydrocarbon Criteria Working Group Methodology would not have been possible without the hard work and dedication of individuals and organizations across the public and private sectors. Specifically, we would like to acknowledge the following members of the TPHCWG steering committee:

Beth Albertson*, *Friedman and Bruya*

Roger Andes and Christopher Barkan, *Association of American Railroads*

Bruce Bauman, Roger Claff, Judith Shaw, and Lorraine Twerdock, *API*

Barbara Brooks, *Hawaii Department of Health*

Deborah Edwards and Joan Tell, *Exxon Biomedical Sciences, Inc.*

John Fitzgerald, *Massachusetts Department of Environmental Protection*

Kathy Garland*, *New Mexico EMNRD*

Joseph Greenblott and Bruce Peirano, *US EPA*

John Gustafson, Bruce Krewinghaus, Ross MacDonald, Ileana Rhodes, *Shell Development Company*

Keith Hoddinott, *US Army*

Jim Holler, *Agency for Toxic Substances and Disease Registry*

Paul Kostecki and Tom Potter, *University of Massachusetts, Amherst*

William Kucharski*, *GECCO Inc.*

Katherine Kurtz, *Navy Environmental Health Center*

Mark Laughman, *Louisiana Department of Environmental Quality*

David Linz*, *LinzTech*

David Nakles, *RETEC*

Doug Orem and Susan Youngren, *EA Environmental Science and Engineering*

Wade Weisman*, *US Air Force Armstrong Lab*

Robert Wilkenfeld*, *Chevron*

* members of Executive Committee

The TPHCWG would like to thank the following reviewers for their helpful suggestions and comments: Lynn Coleman, Washington Department of Ecology; John Thompson, Fluor Daniel GTI; Peter Miasek, Imperial Oil; Bill Lowery, NJ DEP; Adolfo Silva, Petro-Canada; Roger Brewer, Dames & Moore; Fred Reitman and True-Jenn Sun, Texaco; Paul Kuhlmeier; Hans Stroo, RETEC; Victor Kremesec, Amoco; Paul Zahn, Utah DEQ; and Daniel Smith, USAF.

Additionally, we would like to commend Donna Voorhees (Menzie-Cura) and Tamlyn Oliver (AEHS) for their adept assistance in editing and publication of this document.

1.0 INTRODUCTION

1.1 DESCRIPTION

Many exposure pathways which need to be evaluated for petroleum hydrocarbon contaminated soils are cross-media in nature (e.g., leaching of chemicals in soils to groundwater) and thus require analysis of chemical fate and transport in their evaluation. Most current approaches use analytical models to estimate the movement of the chemical from the source to a receptor. At present, these methods model the transport of chemicals in the environment individually. This approach is appropriate for sites that focus on a relatively small number of chemicals. However, for sites containing large numbers of individual compounds (such as petroleum hydrocarbon distillates and crude oils) this method is inadequate, due primarily to excessive analytical and computational requirements; and insufficient fate and transport and toxicological data.

The Total Petroleum Hydrocarbon Criteria Working Group (the "Working Group") convened to provide a solution to this problem. This volume describes how the Working Group grouped petroleum hydrocarbons into a relatively small number of fractions with similar physical-chemical properties, simplifying modeling of their movement in the environment. Fraction-specific properties can then be used to estimate the partitioning of the specific fraction in soil-water-air systems. Fate and transport models (either simple or complex) can then be applied as well.[1]

More than 200 hydrocarbons were considered in the development of fraction-specific properties. A simple screening-level, partitioning model based on the ASTM Standard Guide for Risk-Based Corrective Action Applied at Petroleum Release Sites, "RBCA" (ASTM, 1995) was applied to each chemical in order to quantify, individually, the chemical's relative ability to leach from soil to groundwater and volatilize from soil to air. Based on the modeling results, the chemicals were grouped into fractions (using an order of magnitude as the cutoff point). Once the fractions were defined, typical fate and transport properties were assigned to each fraction based on an empirical relationship between fate and transport properties of chemicals within each fraction and boiling point. These properties could be used to estimate fraction-specific exposure potential at petroleum hydrocarbon contaminated sites.

[1] Although this report does not specifically address biodegradation, it is an important fate process when estimating exposure and should be considered in site-specific analyses.

1.2 RISK-BASED APPROACH TO EVALUATING TPH CONTAMINATED SITES

The fate and transport of a chemical defines its exposure potential. The combination of chemical properties, site properties, and information on how individuals or species (called receptors) are potentially placed in contact with the chemical all define the Exposure Scenario. Estimation of chemical concentrations at a receptor location is an Exposure Assessment. A route of exposure is defined which describes the way receptor to chemical contact could occur (e.g., inhalation, ingestion).

Exposure scenario information and assumptions are combined with chemical toxicity information to calculate potential human health risk. The same equations used to calculate potential human health risk can be used to back-calculate environmental media concentrations corresponding to an acceptable risk level. This risk-based approach to calculating soil cleanup levels, or risk-based screening levels (RBSLs), is recommended by the American Society for Testing and Materials (ASTM) in its Risk-Based Corrective Action (RBCA) Framework (1995) and by the U.S. Environmental Protection Agency (U.S. EPA) in its Soil Screening Guidance (U.S. EPA, 1996). Using the RBCA and EPA frameworks (but not necessarily the prescribed equations), the Working Group has established an approach for estimating RBSLs for TPH at contaminated sites.

1.3 A FRACTION APPROACH FOR COMPLEX MIXTURES

Because data are unavailable for many of the individual components of petroleum hydrocarbons, we have characterized typical physical-chemical properties for subsets of an entire mixture (called fractions) using available data from the literature. Fractions are determined based on a range of physical-chemical properties and simple partitioning models. It is important not to treat a mixture as behaving in the environment as a single, pure liquid. Hydrocarbon mixtures separate and partition based on the makeup of the individual chemicals. Some chemicals persist, some degrade, some are mobile, some adhere tightly to soil particles. However, it is reasonable to assume that chemicals of similar nature (aliphatic or aromatic) and boiling point would behave similarly. We have specified the delineation of the specific fractions based on an order of magnitude differentiation in these simple partitioning properties (i.e., leaching to groundwater and volatilization to air).

Once the fate and transport fractions are specified, fraction-specific toxicity values (reference dose, RfD and reference concentration, RfC) for a non-cancer endpoint[2] were determined for each fraction by the Working Group's Toxicology Technical Advisory Committee. As toxicity data are even less available for complex mixtures, the same toxicology value could be assigned to different fate and transport fractions. However, it is important that the fractions remain differentiated so that the exposure potential estimation remains intact. Fate and transport fractions with their corresponding RfD and RfC are then used to estimate fraction-specific RBSLs for different exposure scenarios.

These RBSLs are protective for non-cancer human health effects. Potential carcinogens must be evaluated separately. This screening level analysis would signif-

[2] Risks due to carcinogenic exposures of chemicals should be handled individually, using an indicator approach.

icantly simplify the analysis of complex hydrocarbon mixtures at petroleum release sites. Furthermore, the individual RBSLs for each fraction can be combined with a site-specific mixture composition to compute a single mixture-specific TPH RBSL value.

1.4 ANALYTICAL REQUIREMENTS FOR FRACTIONS

The definition of the fractions must be consistent with available analytical capabilities. The mass percent of each fraction in the whole mixture or a mixture similar to the type on site is needed when calculating a mixture-specific TPH concentration. The concentration of each fraction in the soil is important to determining which fractions are driving the risk on a site (and concomitantly, which fractions pose no threat). Generic or typical mixture compositions based on historical site information could be used, or distinct analysis applied to site samples. Analytical approaches are published separately.

2.0 MIGRATION PATHWAYS

2.1 PARTITIONING BEHAVIOR

The simple partitioning of a chemical or group of chemicals (i.e., fractions) can be determined using mass balance relationships described in Feenstra et al. (1991). These summary equations are presented below. Depending on the chemical and basic soil properties, a chemical profile can be developed which relays information on where the chemical will reside in the soil.

$$\text{Water Phase Mass Fraction} = \frac{\rho_s}{\theta_{ws} + k_s\rho_s + H\theta_{as}} \quad (1)$$

$$\text{Vapor Phase Mass Fraction} = \frac{H\rho_s}{\theta_{ws} + k_s\rho_s + H\theta_{as}} \quad (2)$$

$$\text{Sorbed Phase Mass Fraction} = \frac{k_s\rho_s}{\theta_{ws} + k_s\rho_s + H\theta_{as}} \quad (3)$$

where:

ρ_s = soil bulk density [g/cm^3]

θ_{ws} = volumetric water content in vadose zone soils [cm^3/cm^3]

k_s = soil-water sorption coefficient [cm^3/g]
(k_s is a function of organic carbon partition coefficient and organic carbon soil content)

H = Henry's law constant [cm^3/cm^3]

θ_{as} = volumetric air content in vadose-zone soils [cm^3/cm^3].

The total of the mass fractions should equal unity if no degradation or other losses are apparent. Therefore, this is a simple way to compare the theoretical disposition of a chemical or fraction in the environment.

2.2 TRANSPORT EQUATIONS

Two predominant transport mechanisms that are directed by a chemical's partitioning in the sub-surface soil are leaching to groundwater and volatilization to air. There are several methods available to characterize the leaching or volatilization potential of a chemical (see also ASTM, 1995; U.S. EPA, 1996). For this project we have relied on the simple screening-level models presented in the ASTM RBCA standard. These models capture the fundamental mechanisms of contaminant migration (e.g., aqueous phase convection and vapor phase diffusion) without requiring excessive computation. With this approach, several key simplifying assumptions are made. They include:

- No degradation (biotic or abiotic) of the chemical in soil or groundwater
- Simple, linear equilibrium partitioning of the chemical between soil particles and soil water.
- Uniform soil moisture content with depth (until the capillary fringe)
- No additional dilution of the chemical occurs once it reaches the groundwater table (no lateral dispersion)
- No free product (NAPL) is present.[3]

The ASTM RBCA framework provides methods for estimating concentrations in air, water, and soil which pose minimal health risk based on exposure factors and toxicity measures. Some equations additionally rely on trans-media migration. For example, risk-based screening levels of Chemical A in air ($RBSL_{air}$) may be calculated based on a no-effect concentration and data on how much air an individual breathes in a day, body weight, and exposure duration, etc. Alternately, air concentrations may also be based on other criteria such as explosive hazards (LEL) or aesthetic criteria such as odor thresholds. Risk-based screening levels for the volatilization to air pathway ($RBSL_{soil}$) are then estimated by dividing the screening level concentration in air (mg/m^3) by a trans-media (soil to air) partitioning value, the volatilization factor (VF).

$$RBSL_{soil} = \frac{RBSL_{air}}{VF} \qquad (4)$$

[3] Although this report does not specifically address nonaqueous phase liquid (NAPL) contamination, this approach is not invalid for sites containing NAPLs. For cases where NAPL is present, the upper limit in concentration at saturation in soil vapor and soil moisture must be considered and models relevant to free-phase products applied.

Similarly, risk-based soil concentrations are estimated for the leaching-to-groundwater pathway, where the "safe" drinking water concentration (mg/L) is divided by a soil-to-water trans-media partition value, the leaching factor (LF). This type of multi-pathway approach is consistent with current EPA methods (U.S. EPA, 1996).

$$RBSL_{soil} = \frac{RBSL_{water}}{LF} \qquad (5)$$

The leaching to groundwater pathway is analytically described by the Leaching Factor. The leaching factor (Equation 6) can be characterized as the ratio of the chemical concentration in groundwater to the chemical concentration in the subsurface soil. This equation does not consider any dilution or attenuation of the chemical in the vadose zone. Essentially, it is applicable to cases where contaminated vadose zone soil comes in close contact with the underlying groundwater. However, it is a useful parameter for characterizing the migration potential of a chemical given a set of specific soil properties.

$$LF = \frac{\rho_s}{[\theta_{ws} + k_s\rho_s + H\theta_{as}](1 + \frac{U_{gw}\delta_{gw}}{IW})} \qquad (6)$$

where:

LF = leaching factor [mg/L-H_2O/mg/kg-soil]

U_{gw} = groundwater Darcy velocity [cm/y]

δ_{gw} = groundwater mixing zone thickness [cm]

I = infiltration rate of water through soil [cm/y]

W = width of source area parallel to groundwater flow direction [cm]

There are several volatilization factor (VF) equations presented in the RBCA standard (volatilization from surface soils, subsurface soils, and into enclosed spaces). The equations are similar, relying on similar chemical properties as input parameters, and would result in relatively, similar values. For this analysis, only one equation is needed and results would not differ significantly if alternate equations were used. Thus, the subsurface soil volatilization factor was selected as the descriptor of chemical transport to air (Equation 7). In this case, the volatilization factor can be described as the ratio between chemical concentration in air and chemical concentration in subsurface soil.

$$VF = \frac{H\rho_s \times 10^3}{[\theta_{ws} + k_s\rho_s + H\theta_{as}](1 + \frac{U_{air}\delta_{air}L_s}{D_s^{eff}W})} \qquad (7)$$

where:

VF = volatilization factor [mg/m^3-air/mg/kg soil]

U_{air} = wind speed above ground surface in ambient mixing zone [cm/s]

δ_{air} = ambient air mixing zone height [cm]

L_s = depth to subsurface soil sources [cm]

D_s^{eff} = effective diffusion coefficient between groundwater and soil surface [cm^2/s]

W = width of source area parallel to wind direction [cm]

The effective diffusion coefficient is further defined by Millington and Quirk (1961) as:

$$D_s^{eff} = D_{air} \frac{\theta_{as}^{3.33}}{\theta_t^2} + D_w \frac{1}{H} \frac{\theta_{ws}^{3.33}}{\theta_t^2} \qquad (8)$$

where:

D_{air} = diffusivity in air [cm^2/s]

θ_t = total soil porosity [cm^3/cm^3]

D_w = diffusivity in water [cm^2/s]

Input parameters can be divided into two parts: site parameters and chemical-specific parameters. For this generic analysis in which we are focusing on chemical properties, site parameters were held constant across all analyses at the default values used in the ASTM RBCA standard (listed in Table 5). In Section 4 we describe how these transport equations were used to select TPH fractions and to develop physical and chemical properties for each fraction.

2.3 SOIL SATURATION

In the environment, where soil, water, and air interact with chemicals, the maximum concentration of a chemical in soil is limited by saturation. This soil saturation concentration, C_{sat}, corresponds to the chemical concentration in soil at which sorption limits of the soil particles, solubility limits of the soil pore water, and saturation of soil pore air have been reached (U.S. EPA, 1996). That is, solubility and vapor pressure limits place an upper limit on the potential for exposure (and thus risk). From a health risk assessment standpoint, if risk levels are acceptable at C_{sat} for pathways which consider trans-media partitioning, then concentrations greater than C_{sat} would also be acceptable because the flux of the chemical from the soil to the soil air or soil water reaches a plateau. Emissions (and thus exposure or health risk) will not increase above the level corresponding to C_{sat} no matter how much more chemical is added to the soil. The equation for C_{sat} is:

$$C_{sat} = \frac{S}{\rho_s} \times [H\theta_{as} + \theta_{ws} + k_s\rho_s] \qquad (9)$$

where:

C_{sat} = soil saturation limit [mg/kg]

S = water solubility [mg/L]

For hydrocarbon mixtures, the effective solubility of individual components, calculated by Raoult's law, is used to calculate $C_{sat,i}$. Raoult's law governs the composition of the chemical mixture in the vapor and water phases in the soil. Raoult's law states that the concentration of chemicals in the vapor and moisture phases that are in equilibrium with soil at concentrations greater than saturation are functions of the mole fraction of each component (X) in the separate hydrocarbon phase (Equation 10). This will result in calculated C_{sat} values lower than those from Equation 9.

$$C_{sat,i} = S_i\, X_i \frac{(\theta_{as} H_i + \theta_{ws} + k_{s,i}\rho_s)}{\rho_s} \qquad (10)$$

where:

X = mole fraction of component i [moles i/moles total]

i = indicates values for individual chemicals

While saturation does provide an upper limit to the vapor and moisture phase concentrations, C_{sat} is only a conservative indication of where limits of solubility and vapor pressure will be reached for a chemical It is not an indication of when residual or free-phase hydrocarbon (NAPL) will be present.

C_{sat} is a delineation point for the applicability of the simple screening models used in this methodology. Thus, the partitioning equations presented here are not appropriate for soils containing free product (primarily due to limits of applicability of the Henry's law constant). For sites where NAPL is present it is also important to determine if the product is mobile. Mobility of an immiscible liquid such as petroleum in soil depends on a balance of capillary, gravitational, hydrodynamic, and surface tension forces. In unsaturated soils, the residual (immobile) concentration of product ranges from 5% to 20% of total pore volume, while in the saturated zone these concentrations are higher, with typical values ranging from 15% to 50% of total pore volume (U.S. EPA, 1992). Therefore, C_{sat} should not be used as a *de facto* RBSL if calculated cleanup goals are greater than C_{sat}. Cases where RBSLs are greater than C_{sat} indicate that no amount of a chemical will pose a human health risk by a particular cross-media exposure pathway and risk managers may consider other criteria for determining cleanup goals (e.g., ecological end points, aesthetic criteria, etc.).

3.0 PHYSICAL AND CHEMICAL PROPERTIES OF HYDROCARBON COMPOUNDS

3.1 GENERAL PROPERTIES

As discussed briefly in the previous section, several chemical properties are required to evaluate how a chemical partitions in tri-phasic (air-water-soil) systems. Table 1 provides a listing of the chemical properties required to estimate leaching and volatilization factors. Some properties are used in supplemental equations. Other properties not used directly in these analyses that may be important in other

Table 1. Chemical Properties

Parameter	Symbol	Units
Molecular weight	MW	g/mole
Aqueous solubility	S	mg/L
Vapor pressure	VP	atm
Henry's law constant	H	atm-m^3/mole or cm^3/cm^3
Boiling point	BP	^{o}C
Equivalent carbon number	EC	unitless
Soil-water sorption coefficient	K_s	cm^3/g
Octanol-water partition coefficient	K_{ow}[a]	cm^3/cm^3
Organic-carbon partition coefficient	K_{oc}	cm^3/g
Diffusivity in air	D_{air}	cm^2/s
Diffusivity in water	D_w	cm^2/s

[a] K_{ow} is also commonly labeled as P.

transport modeling are the specific gravity and melting point. Some of the more important properties are discussed below. As shown in Equation 6, the most sensitive properties used to estimate the leaching factor are the soil-water sorption coefficient, and the Henry's law constant. Likewise, the same properties in addition to diffusivity in air are important when calculating volatilization from subsurface soils to air (diffusivity in water is also used in the latter equation, but it is sufficiently low to render negligible the term in the summation in Equation 8).

A very sensitive parameter not discussed in detail is the biodegradation rate of the chemical. Often the degradation rate will affect the downgradient concentrations by several orders of magnitude and can drive decision-making at a site. However, biodegradation, in addition to being a function of the chemical structure is also a function of the site parameters (soil types, dissolved oxygen content, et al.). Biodegradation rates for a chemical can vary significantly from site to site. A typical approach for handling biodegradation losses at a site is to assume biodegradation = 0 for the initial screening level analyses (Tier I) and then obtain relevant literature data or site-specific data for more detailed iterations (or tiers) in the risk assessment process. It is not in the current scope of the Working Group to develop fraction-specific degradation rates, although this is an apparent research need.

3.1.1 Equivalent Carbon Number, EC

The Equivalent Carbon Number, EC, is related to the boiling point of a chemical normalized to the boiling point of the n-alkanes or its retention time in a boiling point gas chromatographic (GC) column. This relationship, displayed in Figure 1, was empirically determined. Thus, for chemicals where only boiling points are known, an equivalent carbon number can be easily calculated. For example, hexane contains six carbons and has a boiling point of 69°C. Its equivalent carbon number is six. Benzene, also containing six carbons, has a boiling point of 80°C. Based on benzene's boiling point and its retention time in a boiling point GC column, benzene's equivalent carbon number is 6.5. This approach is consistent with methods routinely used in the petroleum industry for separating complex mixtures and is a more appropriate differentiation technique than the carbon number of the chemical. It is typically how analytical laboratories report carbon numbers for chemicals evaluated on a boiling point GC column. Figure 2 displays the relationship between carbon number and equivalent carbon number. Note that for molecules with higher relative carbon number indices, the disparity between aliphatic and aromatic hydrocarbons is great.

3.1.2 Soil-Water Sorption Coefficient, K_s

K_s, the soil-water sorption coefficient, expresses the tendency of a compound to be adsorbed onto a soil or sediment particle. K_s can be measured as the ratio of the amount of the chemical sorbed per unit weight of soil or sediment to the concentration of the chemical in solution. Sorption of a chemical to soil is typically measured in soil-water batch systems where a known concentration of chemical in water is introduced to known quantities of "clean" soil. Over time, the chemical will migrate from the free water and sorb onto the soil particles. The magnitude of the

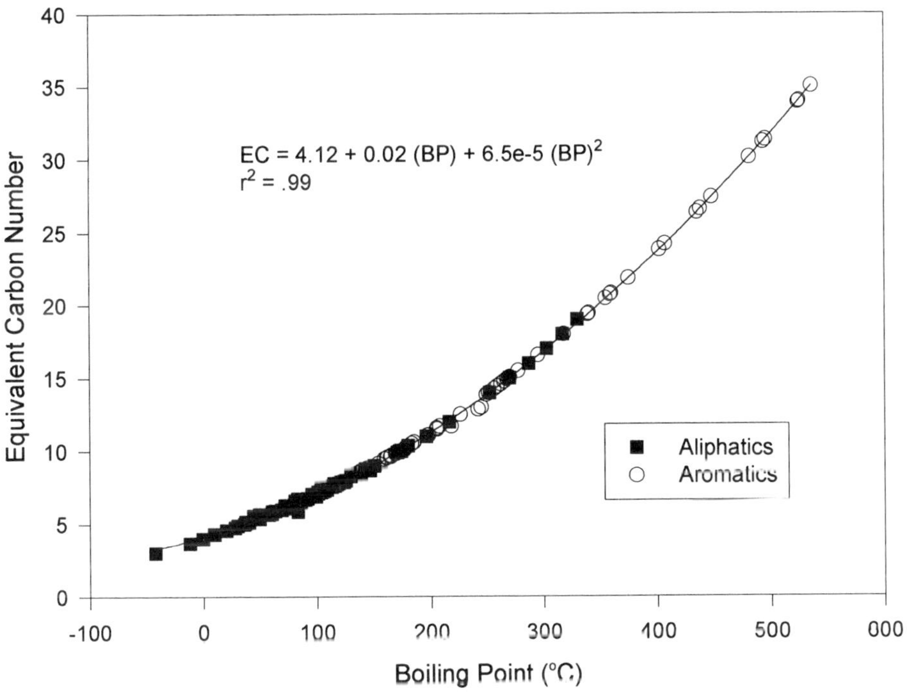

Figure 1. Equivalent Carbon Number: Boiling Point Normalized to n-Alkanes

sorption for most soil-water systems is a function of the hydrophobicity of the chemical (measured by the water solubility) and the organic carbon content of the soil. Once the system reaches chemical equilibrium, concentrations are measured in both the liquid phase and solid phase. The slope of the line plotting the liquid phase concentration vs. the solid phase concentration for various combinations of soil mass and liquid phase concentrations is K_s. There are several assumptions inherent in using this approach:

Sorption of Chemical "A" to Soil "B" is Linear: This method assumes that there is no limit to the number of binding sites on the soil particles and that there is no limit to chemical size in fitting onto binding sites. Studies have shown that sorption processes are decidedly nonlinear for high solute concentrations, ionized chemicals, and metals (Voice and Weber, 1983). However, for non-ionic, hydrophobic chemicals such as hydrocarbons with concentrations significantly less than the solubility limit, the linear assumption is reasonable. Additionally, there is the implicit assumption that the contact time between the chemical and soil is adequate to reach chemical equilibrium.

Implicit in the linear sorption assumption is that the rate of adsorption equals the rate of desorption. Field and laboratory studies have shown that for many chemicals and soil types, the rate of desorption could be significantly lower than the adsorption rate (DiToro, 1985).

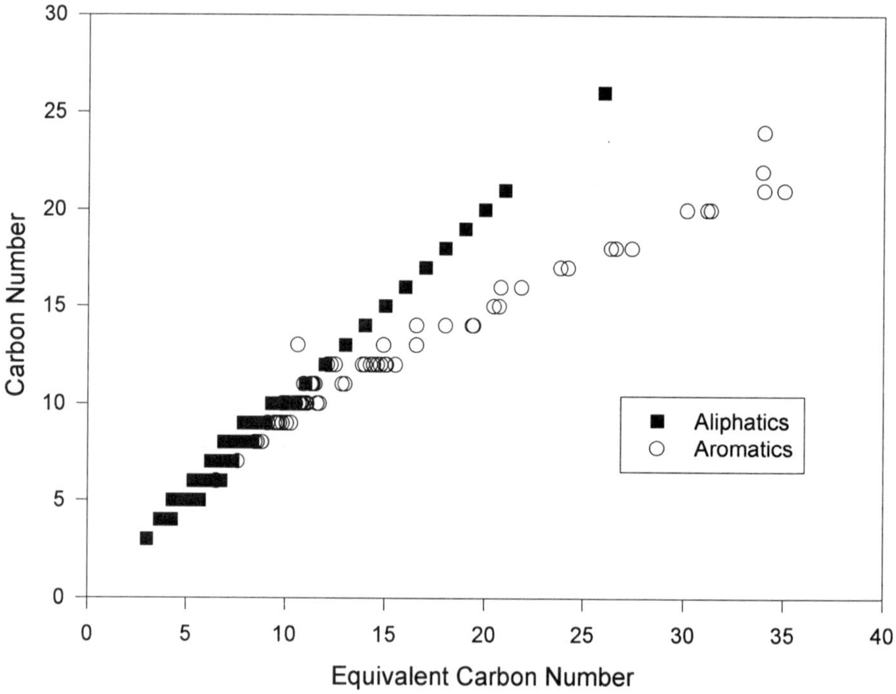

Figure 2. Equivalent Carbon Number vs. Carbon Number

Chemical "A" Binds to Soil "B" Equally Regardless of Soil Moisture Content: Most chemical-specific sorption coefficients found in the literature have been developed using bi-phasic systems (soil-water). In these cases, the soil is completely saturated, leaving no air spaces (designed to mimic groundwater systems). However, K_s values developed for groundwater systems are routinely applied to unsaturated soils when modeling. Differences are probably not significant in tri-phasic systems where the soil moisture content approaches the soil porosity. In cases where the soil moisture content is low, a different sorption profile will be evidenced, with the chemical Henry's law constant becoming a key parameter. This problem is not sufficiently addressed in current screening-level models (or even some more "robust" models). Soil batch studies and column studies can be performed to better quantify this phenomenon.

Organic Carbon Content Is the Controlling Variable: The approaches discussed here assume that a chemical contaminant in soil will bind or sorb to the organic matter of the soil (see discussion below). For soils with organic carbon fractions less than .001, other components of the soil matrix (e.g., inorganic clay minerals) express greater influence. This is also the case for some chemicals with charged functional groups.

In addition to literature values and extrapolations from other variables (see Section 3.1.3), K_s can be measured for site-specific analyses (usually in Tier III analyses). Representative soil samples may be collected at the site and, using solutes typical of site conditions, isotherms developed and soil-column breakthrough curves measured. These results will give a more realistic picture of chemical binding to site soils. However, care should be taken as soils can vary across a site and with depth. For this analysis, K_s was calculated by multiplying the organic carbon partition coefficient K_{oc} (which is discussed in Sections 3.1.3 and 3.2.2.2) by a typical value of the fraction of organic carbon f_{oc} (see Table 5).

3.1.3 Organic-Carbon Partition Coefficient, K_{oc}

Researchers have found that for non-ionic hydrophobic chemicals, the primary soil property controlling sorption is the organic carbon content of the soil (Karickhoff et al., 1979). This is assumed to be a linear relationship. In order to compare studies of the same chemical on different soils, one can calculate the organic-carbon partition coefficient, K_{oc}. K_{oc} is defined by the following equation:

$$K_{oc} = \frac{K_s}{f_{oc}} \qquad (11)$$

with units of:

$$\frac{\left[\dfrac{\left(\text{mg chem}/\text{kg soil}\right)}{\left(\text{mg chem}/\text{L water}\right)}\right]}{\left[\text{kg o.c.}/\text{kg soil}\right]} = \frac{\text{L water}}{\text{kg o.c.}}$$

where:

f_{oc} = fraction of organic carbon [kg organic carbon/kg soil].

K_{oc} values may range from 1 to 10,000,000 cm^3/g (Lyman et al., 1992) and are usually presented in the log10 form. Because of the variability in the measured and estimated K_{oc} values found in the literature for some compounds, and lack of available data for many compounds, K_{oc} in this study was estimated directly from the octanol-water partition coefficient, K_{ow}. Lyman et al. (1990) presents a summary of equations that calculate log K_{oc} based on other more readily available parameters (i.e., K_{ow}). The method of estimation used is discussed in Section 3.2.2. The fraction of organic carbon, f_{oc}, is a function of soil type and depth. Generally, soil organic carbon is about 60% of the total soil organic matter content (Hillel, 1980). Typical values of soil organic matter for sandy-loam soils range from approximately 1% to 4%. Sands and clay soils have less organic carbon (<1%) and loams have greater (up to 10%). Peat soils could have up to 50% organic matter. Organic matter content decreases with depth.

3.1.4 Octanol-Water Partition Coefficient, K_{ow}

K_{ow}, the octanol-water partition coefficient is defined as the ratio of the chemical's concentration in the octanol phase to its concentration in the aqueous phase of a two-phase octanol-water system.

$$K_{ow} = \frac{\text{concentration in octanol phase [mg/L]}}{\text{concentration in aqueous phase [mg/L]}} \qquad (12)$$

K_{ow} provides an estimated value of a chemical's hydrophobicity or partitioning tendency from water to organic matter, and thus can be correlated to the compound's water solubility and K_{oc}. Several estimation methods are available for estimating K_{ow} when data are lacking (see Section 3.2.2).

3.1.5 Henry's Law Constant, H

The Henry's law constant, H, can be defined as an air-water partition coefficient and can be measured as the ratio of a compound's concentration in air to its concentration in water at equilibrium.

$$H = \frac{\text{concentration in air [atm]}}{\text{concentration in water [mole/m}^3\text{]}} \qquad (13)$$

H is calculated by taking the ratio of vapor pressure to aqueous solubility, and has units of atm-m^3/mole. H is often expressed in its concentration-based form (units of cm^3 water/cm^3 air) by dividing the parameterized form of H by the universal gas constant ($R=8.2x10^{-5}$ atm-m^3/mole-K) and temperature, T (in degrees K). The Henry's law relationship is only valid for dilute solutions where water concentrations are less than the water solubility. When petroleum concentrations in soils and soil-water approach residual saturation, this relationship may not be valid.

3.1.6 Diffusivity in Air, D_{air}

The diffusion coefficient or diffusivity in air, D_{air}, is a measure of the diffusion of a molecule in a gas medium as a result of intermolecular collisions. It is not a measure of turbulence or bulk transport (apparent diffusion). Diffusivity is technically defined as:

$$D_{air-B} = \frac{J_B}{\nabla X_B} \quad (14)$$

where:

D_{air-B} = diffusivity of compound B in compound or mixture A (in this case A is air) [cm^2/s]

J_B = net molal flux of B [mol/cm^2-s]

∇X_B = concentration gradient of B [mol/cm^3-cm]

Most environmental fluid media are turbulent in nature and therefore controlled by the intensity of turbulent mixing rather than molecular diffusion. Situations where molecular diffusion is significant, or even controlling, include the movement of chemicals at air-water interfaces, the interstitial waters of sediments, and groundwater (Lyman et al., 1990). However, hydrodynamic and mechanical dispersion plays a major role in determining chemical concentrations at a receptor (mixing zone effect), a parameter that is not considered in these simple screening-level models.

3.1.7 Diffusivity in Water, D_{wat}

The diffusion coefficient in water, D_{wat}, is a function of solute size, temperature, and solution viscosity. D_{wat} is not a sensitive parameter in this analysis; however, the equation is presented below for completeness.

$$D_{wat-B} = \frac{R\,T}{6\,\eta_W\,r_B} \quad (15)$$

where:

D_{wat-B} = diffusivity of compound B in water [cm^2/s]

η_W = viscosity of water [cp]

r_B = radius of molecule B [cm]

3.2 CONSTITUENT HYDROCARBON COMPOUNDS

A literature search was performed by the Working Group's Analytical Technical Advisory Committee to compile a list of compounds found in petroleum distillate products and crude oils and their relevant physical-chemical properties. Table 2 presents the compounds and their relative abundance in specific petroleum products: gasoline, crude oil, JP-4, JP-5, JP-8, kerosene, diesel, and home heating oil (when available). TPHCWG Volume 2 contains additional composition information.

Table 2. Petroleum Hydrocarbon Composition of Fuels

Compound	Number of Carbons	EC	Weight Percent	Fuel Type	Reference
STRAIGHT CHAIN ALKANES					
Propane	3	3	0.01 - 0.14	Gasoline	LUFT, 1988
n-Butane	4	4	3.93 - 4.70	Gasoline	LUFT, 1988
			0.12	JP-4	API, 1993
n-Pentane	5	5	5.75 - 10.92	Gasoline	LUFT, 1988
			1.06	JP-4	API, 1993
n-Hexane	6	6	0.24 - 3.50	Gasoline	LUFT, 1988
			0.7 - 1.8	Crude Oil	API, 1993
			2.21	JP-4	API, 1993
n-Heptane	7	7	0.31 - 1.96	Gasoline	LUFT, 1988
			0.8 - 2.3	Crude Oil	API, 1993
			3.67	JP-4	API, 1993
			0.03	JP-8	API, 1993
			0.1	Kerosene	API, 1993
n-Octane	8	8	0.36 - 1.43	Gasoline	LUFT, 1988
			0.9 - 1.9	Crude Oil	API, 1993
			3.8	JP-4	API, 1993
			0.12	JP-5	API, 1993
			0.9	JP-8	API, 1993
			0.2 - 0.3	Kerosene	API, 1993
			0.1	Diesel	BP, 1996
			0.1	Fuel Oil #2	BP, 1996
n-Nonane	9	9	0.07 - 0.83	Gasoline	LUFT, 1988
			0.6 - 1.9	Crude Oil	API, 1993
			2.25	JP-4	API, 1993
			0.38	JP-5	API, 1993
			0.31	JP-8	API, 1993
			0.4 - 0.8	Kerosene	API, 1993
			0.19 - 0.49	Diesel	BP, 1996
			0.20 - 0.30	Fuel Oil #2	BP, 1996

Table 2. *Continued*

Compound	Number of Carbons	EC	Weight Percent	Fuel Type	Reference
n-Decane	10	10	0.04 - 0.50	Gasoline	LUFT, 1988
			1.8	Crude Oil	API, 1993
			2.16	JP-4	API, 1993
			1.79	JP-5	API, 1993
			1.31	JP-8	API, 1993
			1.5 - 1.7	Kerosene	API, 1993
			0.28 - 1.2	Diesel	BP, 1996
			0.5	Fuel Oil #2	BP, 1996
n-Undecane	11	11	0.05 - 0.22	Gasoline	LUFT, 1988
			1.7	Crude Oil	API, 1993
			2.32	JP-4	API, 1993
			3.95	JP-5	API, 1993
			4.13	JP-8	API, 1993
			3.5 - 6.1	Kerosene	API, 1993
			0.57 - 2.3	Diesel	BP, 1996
			0.80 - 0.90	Fuel Oil #2	BP, 1996
n-Dodecane	12	12	0.04 - 0.09	Gasoline	LUFT, 1988
			1.7	Crude Oil	API, 1993
			2	JP-4	API, 1993
			3.94	JP-5	API, 1993
			4.72	JP-8	API, 1993
			2.8 - 5.7	Kerosene	API, 1993
			1.0 - 2.5	Diesel	BP, 1996
			0.84 - 1.20	Fuel Oil #2	BP, 1996
n-Tridecane	13	13	1.52	JP-4	API, 1993
			3.45	JP-5	API, 1993
			4.43	JP-8	API, 1993
			3.1 - 5.2	Kerosene	API, 1993
			1.5 - 2.8	Diesel	BP, 1996
			0.96 - 2.00	Fuel Oil #2	BP, 1996

Table 2. *Continued*

Compound	Number of Carbons	EC	Weight Percent	Fuel Type	Reference
n-Tetradecane	14	14	0.73	JP-4	API, 1993
			2.72	JP-5	API, 1993
			2.99	JP-8	API, 1993
			2.3 - 4.7	Kerosene	API, 1993
			0.61 - 2.7	Diesel	BP, 1996
			1.03 - 2.50	Fuel Oil #2	BP, 1996
n-Pentadecane	15	15	1.67	JP-5	API, 1993
			1.61	JP-8	API, 1993
			0.6 - 2.3	Kerosene	API, 1993
			1.9 - 3.1	Diesel	BP, 1996
			1.13 - 3.20	Fuel Oil #2	BP, 1996
n-Hexadecane	16	16	1.07	JP-5	API, 1993
			0.45	JP-8	API, 1993
			0.1 - 0.7	Kerosene	API, 1993
			1.5 - 2.8	Diesel	BP, 1996
			1.05 - 3.30	Fuel Oil #2	BP, 1996
n-Heptadecane	17	17	0.12	JP-5	API, 1993
			0.08	JP-8	API, 1993
			0.4	Kerosene	API, 1993
			1.4 - 2.9	Diesel	BP, 1996
			0.65 - 3.60	Fuel Oil #2	BP, 1996
n-Octadecane	18	18	0.02	JP-8	API, 1993
			0.3	Kerosene	API, 1993
			1.2 - 2.0	Diesel	BP, 1996
			0.55 - 2.50	Fuel Oil #2	BP, 1996
n-Nonadecane	19	19	0.2	Kerosene	API, 1993
			0.7 - 1.5	Diesel	BP, 1996
			0.33 - 1.30	Fuel Oil #2	BP, 1996
n-Eicosane	20	20	0.1	Kerosene	API, 1993
			0.4 - 1.0	Diesel	BP, 1996
			0.18 - 0.60	Fuel Oil #2	BP, 1996

Table 2. *Continued*

Compound	Number of Carbons	EC	Weight Percent	Fuel Type	Reference
n-Heneicosane	21	21	0.1	Kerosene	API, 1993
			0.26 - 0.83	Diesel	BP, 1996
			0.09 - 0.40	Fuel Oil #2	BP, 1996
n-Docosane	22	22	0.14 - 0.44	Diesel	BP, 1996
			0.1	Fuel Oil #2	BP, 1996
n-Tetracosane	24	24	0.35	Diesel	BP, 1996
n-Hexacosane	26	26			
BRANCHED CHAIN ALKANES					
Isobutane	4	3.67	0.12 - 0.37	Gasoline	LUFT, 1988
			0.66	JP-4	API, 1993
2,2-Dimethylbutane	6	5.37	0.17 - 0.84	Gasoline	LUFT, 1988
			0.04	Crude Oil	API, 1993
			0.1	JP-4	API, 1993
2,3-Dimethylbutane	6	5.68	0.59 - 1.55	Gasoline	LUFT, 1988
			0.04 - 0.14	Crude Oil	API, 1993
2,2,3-Trimethylbutane	7	6.36	0.01 - 0.04	Gasoline	LUFT, 1988
2,2,3,3-Tetramethylbutane	8	7.3	0.24	JP-4	API, 1993
Neopentane	5	4.32	0.02 - 0.05	Gasoline	LUFT, 1988
Isopentane	5	4.75	6.07 - 10.17	Gasoline	LUFT, 1988
2-Methylpentane	6	5.72	2.91 - 3.85	Gasoline	LUFT, 1988
			0.3 - 0.4	Crude Oil	API, 1993
			1.28	JP-4	API, 1993
3-Methylpentane	6	5.85	2.4 (vol)	Gasoline	LUFT, 1988
			0.3 - 0.4	Crude Oil	API, 1993
			0.89	JP-4	API, 1993
3-Ethylpentane	7		0.05	Crude Oil	API, 1993
2,2-Dimethylpentane	7	6.25	0.25	JP-4	API, 1993
2,4-Dimethylpentane	7	6.31	0.23 - 1.71	Gasoline	LUFT, 1988
			0.05	Crude Oil	API, 1993
2,3-Dimethylpentane	7	6.69	0.32 - 4.17	Gasoline	LUFT, 1988
			0.1 - 0.6	Crude Oil	API, 1993
3,3-Dimethylpentane	7	6.55	0.02 - 0.03	Gasoline	LUFT, 1988

Table 2. *Continued*

Compound	Number of Carbons	EC	Weight Percent	Fuel Type	Reference
2,2,3-Trimethylpentane	8	7.37	0.09 - 0.23	Gasoline	LUFT, 1988
2,2,4-Trimethylpentane	8	6.89	0.32 - 4.58	Gasoline	LUFT, 1988
			0.004	Crude Oil	API, 1993
2,3,3-Trimethylpentane	8	7.58	0.05 - 2.28	Gasoline	LUFT, 1988
			0.006	Crude Oil	API, 1993
2,3,4-Trimethylpentane	8	7.55	0.11 - 2.80	Gasoline	LUFT, 1988
			0.005	Crude Oil	API, 1993
2-Methyl-3-ethylpentane	8	7.66	0.04	Crude Oil	API, 1993
2,4-Dimethyl-3-ethylpentane	9		0.03 - 0.07	Gasoline	LUFT, 1988
2-Methylhexane	7	6.68	0.36 - 1.48	Gasoline	LUFT, 1988
			0.7	Crude Oil	API, 1993
			2.35	JP-4	API, 1993
3-Methylhexane	7	6.76	0.30 - 1.77	Gasoline	LUFT, 1988
			0.19 - 0.5	Crude Oil	API, 1993
			1.97	JP-4	API, 1993
2,2-Dimethylhexane	8	7.25	0.01 - 0.1	Crude Oil	API, 1993
			0.71	JP-4	API, 1993
2,3-Dimethylhexane	8	7.65	0.06 - 0.16	Crude Oil	API, 1993
2,4-Dimethylhexane	8	7.38	0.34 - 0.82	Gasoline	LUFT, 1988
			0.06	Crude Oil	API, 1993
			0.58	JP-4	API, 1993
2,5-Dimethylhexane	8	7.36	0.24 - 0.52	Gasoline	LUFT, 1988
			0.06	Crude Oil	API, 1993
			0.37	JP-4	API, 1993
3,3-Dimethylhexane	8	7.45	0.03	Crude Oil	API, 1993
			0.26	JP-4	API, 1993
3,4-Dimethylhexane	8	7.74	0.16 - 0.37	Gasoline	LUFT, 1988
3-Ethylhexane	8	7.79	0.01	Gasoline	LUFT, 1988
2-Methyl-3-ethylhexane	9		0.04 - 0.13	Gasoline	LUFT, 1988
2,2,4-Trimethylhexane	9	7.93	0.11 - 0.18	Gasoline	LUFT, 1988
2,2,5-Trimethylhexane	9	7.87	0.17 - 5.89	Gasoline	LUFT, 1988
2,3,3-Trimethylhexane	9		0.05 - 0.12	Gasoline	LUFT, 1988

Table 2. *Continued*

Compound	Number of Carbons	EC	Weight Percent	Fuel Type	Reference
2,3,5-Trimethylhexane	9	8.24	0.05 - 1.09	Gasoline	LUFT, 1988
2,4,4-Trimethylhexane	9	8.07	0.02 - 0.16	Gasoline	LUFT, 1988
2-Methylheptane	8	7.71	0.48 - 1.05	Gasoline	LUFT, 1988
			2.7	JP-4	API, 1993
3-Methylheptane	8	7.78	0.63 - 1.54	Gasoline	LUFT, 1988
			3.04	JP-4	API, 1993
4-Methylheptane	8	7.72	0.22 - 0.52	Gasoline	LUFT, 1988
			0.92	JP-4	API, 1993
2,2-Dimethylheptane	9	8.28	0.01 - 0.08	Gasoline	LUFT, 1988
2,3-Dimethylheptane	9	8.64	0.13 - 0.51	Gasoline	LUFT, 1988
			0.05	Crude Oil	API, 1993
2,4-Dimethylheptane	9	8.34	0.43	JP-4	API, 1993
2,5-Dimethylheptane	9	8.47	0.52	JP-4	API, 1993
2,6-Dimethylheptane	9	8.47	0.07 - 0.23	Gasoline	LUFT, 1988
			0.05 -0.25	Crude Oil	API, 1993
3,3-Dimethylheptane	9	8.42	0.01 - 0.08	Gasoline	LUFT, 1988
3,4-Dimethylheptane	9	8.62	0.07 - 0.33	Gasoline	LUFT, 1988
2,2,4-Trimethylheptane	10		0.12 - 1.70	Gasoline	LUFT, 1988
2,4,6-Trimethylheptane	10		0.07	JP-5	API, 1993
			0.07	JP-8	API, 1993
3,3,5-Trimethylheptane	10		0.02 - 0.06	Gasoline	LUFT, 1988
3-Ethylheptane	9	8.77	0.02 - 0.16	Gasoline	LUFT, 1988
4-Ethylheptane	9	8.69	0.18	JP-4	API, 1993
2-Methyloctane	9		0.14 - 0.62	Gasoline	LUFT, 1988
			0.4	Crude Oil	API, 1993
			0.88	JP-4	API, 1993
3-Methyloctane	9	8.78	0.34 - 0.85	Gasoline	LUFT, 1988
			0.1 - 0.4	Crude Oil	API, 1993
			0.79	JP-4	API, 1993
			0.07	JP-5	API, 1993
			0.04	JP-8	API, 1993

Table 2. *Continued*

Compound	Number of Carbons	EC	Weight Percent	Fuel Type	Reference
4-Methyloctane	9	8.71	0.11 - 0.55	Gasoline	LUFT, 1988
			0.1	Crude Oil	API, 1993
			0.86	JP-4	API, 1993
2,6-Dimethyloctane	10	9.32	0.06 - 0.12	Gasoline	LUFT, 1988
2-Methylnonane	10	9.72	0.06 - 0.41	Gasoline	LUFT, 1988
3-Methylnonane	10	9.78	0.06 - 0.32	Gasoline	LUFT, 1988
4-Methylnonane	10		0.04 - 0.26	Gasoline	LUFT, 1988
4-Methyldecane	11		0.78	JP-5	API, 1993
2-Methyldecane	11		0.61	JP-5	API, 1993
			0.41	JP-8	API, 1993
2,6-Dimethyldecane	12		0.72	JP-5	API, 1993
			0.66	JP-8	API, 1993
2-Methylundecane	12		0.64	JP-4	API, 1993
			1.39	JP-5	API, 1993
			1.16	JP-8	API, 1993
3-Methylundecane	12		0.09 - 0.28	Diesel	
2-Methyldodecane	13		0.15 - 0.52	Diesel	
2,6-Dimethylundecane	13		0.71	JP-4	API, 1993
			2	JP-5	API, 1993
			2.06	JP-8	API, 1993
3-Methyltridecane	14		0.13 - 0.30	Diesel	
2-Methyltetradecane	15		0.34 - 0.63	Diesel	
CYCLOALKANES					
Cyclopentane	5	5.66	0.19 - 0.58	Gasoline	LUFT, 1988
			0.05	Crude Oil	API, 1993
Methylcyclopentane	6	6.27	not quantified	Gasoline	LUFT, 1988
			0.3 - 0.9	Crude Oil	API, 1993
			1.16	JP-4	API, 1993
1-Methyl-cis-2-ethylcyclopentane	8		0.06 - 0.11	Gasoline	LUFT, 1988

Table 2. *Continued*

Compound	Number of Carbons	EC	Weight Percent	Fuel Type	Reference
1-Methyl-trans-3-ethylcyclopentane	8		0.06 - 0.12	Gasoline	LUFT, 1988
1,1-Dimethylcyclopentane	7	6.72	0.06 - 0.2	Crude Oil	API, 1993
1-cis-2-Dimethylcyclopentane	7	7.21	0.07 - 0.13	Gasoline	LUFT, 1988
			0.54	JP-4	API, 1993
1-trans-2-Dimethylcyclopentane	7	6.87	0.06 - 0.20	Gasoline	LUFT, 1988
			0.15 - .5	Crude Oil	API, 1993
1-cis-3-Dimethylcyclopentane	7	6.82	0.2	Crude Oil	API, 1993
			0.34	JP-4	API, 1993
1-trans-3-Dimethylcyclopentane	7	6.85	0.2 - 0.9	Crude Oil	API, 1993
			0.36	JP-4	API, 1993
1,1,2-Trimethylcyclopentane	8	7.67	0.06 - 0.11	Gasoline	LUFT, 1988
			0.06	Crude Oil	API, 1993
1,1,3-Trimethylcyclopentane	8	7.25	0.3	Crude Oil	API, 1993
1-trans-2-cis-3-Trimethylcyclopentane	8	7.51	0.01 - 0.25	Gasoline	LUFT, 1988
			0.3 - 0.4	Crude Oil	API, 1993
1-trans-2-cis-4-Trimethylcyclopentane	8		0.03 - 0.16	Gasoline	LUFT, 1988
			0.2	Crude Oil	API, 1993
1-trans-2-trans-4-Trimethylcyclopentane	8	7.19			
Ethylcyclopentane	7	7.34	0.14 - 0.21	Gasoline	LUFT, 1988
			0.26	JP-4	API, 1993
n-Propylcyclopentane	8	7.1	0.01 - 0.06	Gasoline	LUFT, 1988
Isopropylcyclopentane	8		0.01 - 0.02	Gasoline	LUFT, 1988
1-cis-3-Dimethylcyclohexane	8	7.75			API, 1993
			0.42	JP-4	API, 1993
1-trans-2-Dimethylcyclohexane	8	7.94	0.3	Crude Oil	API, 1993
1-trans-3-Dimethylcyclohexane	8	7.99	0.05 - 0.12	Gasoline	LUFT, 1988
1,4-Dimethylcyclohexane	8				
Ethylcyclohexane	8	8.38	0.17 - 0.42	Gasoline	LUFT, 1988
			0.2	Crude Oil	API, 1993

Table 2. *Continued*

Compound	Number of Carbons	EC	Weight Percent	Fuel Type	Reference
Cyclohexane	6	6.59	0.7	Crude Oil	API, 1993
			0.08	Gasoline	API, 1993
			1.24	JP-4	API, 1993
Methylcyclohexane	7	7.22	2.27	JP-4	API, 1993
1-Methyl-2-ethylcyclohexane	9		0.39	JP-4	API, 1993
1-Methyl-3-ethylcyclohexane	9		0.17	JP-4	API, 1993
1-Methyl-4-ethylcyclohexane	9		0.48	JP-5	API, 1993
			0.1	JP-8	API, 1993
1,3,5-Trimethylcyclohexane	9		0.99	JP-4	API, 1993
			0.09	JP-5	API, 1993
			0.06	JP-8	API, 1993
1,1,3-Trimethylcyclohexane	9	8.45	0.48	JP-4	API, 1993
			0.05	JP-5	API, 1993
			0.06	JP-8	API, 1993
n-Butylcyclohexane	10		0.7	JP-4	API, 1993
			0.9	JP-5	API, 1993
			0.74	JP-8	API, 1993
n-Propylcyclohexane	9		0.14	JP-8	API, 1993
Hexylcyclohexane	12		0.93	JP-8	API, 1993
Heptylcyclohexane	13		0.99	JP-5	API, 1993
			1	JP-8	API, 1993
Pentylcyclopentane	10	10.37			
1-trans-2-trans-4-Trimethylcyclohexane	9		0.2	Crude Oil	API, 1993
STRAIGHT CHAIN ALKENES					
Propylene	3				
cis-2-Butene	4	4.25	0.13 - 0.17	Gasoline	LUFT, 1988
trans-2-Butene	4	4.1	0.16 - 0.20	Gasoline	LUFT, 1988
Pentene-1	5	4.89	0.33 - 0.45	Gasoline	LUFT, 1988

Table 2. *Continued*

Compound	Number of Carbons	EC	Weight Percent	Fuel Type	Reference
1-Pentyne	5	5.13			
cis-2-Pentene	5	5.16	0.43 - 0.67	Gasoline	LUFT, 1988
trans-2-Pentene	5	5.08	0.52 - 0.90	Gasoline	LUFT, 1988
1-Hexene	6	5.9			
1-Hexyne	6	6.09			
cis-2-Hexene	6	6.14	0.15 - 0.24	Gasoline	LUFT, 1988
trans-2-Hexene	6	6.05	0.18 - 0.36	Gasoline	LUFT, 1988
cis-3-Hexene	6	6.03	0.11 - 0.13	Gasoline	LUFT, 1988
trans-3-Hexene	6	6.02	0.12 - 0.15	Gasoline	LUFT, 1988
cis-3-Heptene	7	7.01	0.14 - 0.17	Gasoline	LUFT, 1988
trans-2-Heptene	7	7.05	0.06 - 0.10	Gasoline	LUFT, 1988
1-Octene	8	7.89			
1-Nonene	9	8.69			
1-Decene	10	9.91			
Tridecene	13		0.45	JP-5	API, 1993
			0.73	JP-8	API, 1993
BRANCHED CHAIN ALKENES					
2-Methyl-1-butene	5	4.96	0.22 - 0.66	Gasoline	LUFT, 1988
3-Methyl-1-butene	5	4.57	0.08 - 0.12	Gasoline	LUFT, 1988
2-Methyl-2-butene	5	5.21	0.96 - 1.28	Gasoline	LUFT, 1988
2,3-Dimethyl-1-butene	6	5.7	0.08 - 0.10	Gasoline	LUFT, 1988
2-Methyl-1-pentene	6	5.89	0.20 - 0.22	Gasoline	LUFT, 1988
2,3-Dimethyl-1-pentene	7		0.01 - 0.02	Gasoline	LUFT, 1988
2,4-Dimethyl-1-pentene	7	6.48	0.02 - 0.03	Gasoline	LUFT, 1988
4,4-Dimethyl-1-pentene	7		0.60 (vol)	Gasoline	LUFT, 1988
2-Methyl-2-pentene	6	6.07	0.27 - 0.32	Gasoline	LUFT, 1988
3-Methyl-cis-2-pentene	6	6.11	0.35 - 0.45	Gasoline	LUFT, 1988
3-Methyl-trans-2-pentene	6	6.22	0.32 - 0.44	Gasoline	LUFT, 1988
4-Methyl-cis-2-pentene	6	5.69	0.04 - 0.05	Gasoline	LUFT, 1988
4-Methyl-trans-2-pentene	6	5.73	0.08 - 0.30	Gasoline	LUFT, 1988
4,4-Dimethyl-cis-2-pentene	7	6.47	0.02	Gasoline	LUFT, 1988
4,4-Dimethyl-trans-2-pentene	7	6.23	Not quantified	Gasoline	LUFT, 1988
3-Ethyl-2-pentene	7	7.07	0.03 - 0.04	Gasoline	LUFT, 1988

Table 2. *Continued*

Compound	Number of Carbons	EC	Weight Percent	Fuel Type	Reference
CYCLOALKENES					
Cyclopentene	5	5.55	0.12 - 0.18	Gasoline	LUFT, 1988
3-Methylcyclopentene	6	6.1	0.03 - 0.08	Gasoline	LUFT, 1988
Cyclohexene	6	6.74	0.03	Gasoline	LUFT, 1988
ALKYL BENZENES					
Benzene	6	6.5	0.12 - 3.50	Gasoline	LUFT, 1988
			0.04 - 0.4	Crude Oil	API, 1993
			0.5	JP-4	API, 1993
			0.003 - 0.10	Diesel	BP, 1996
			<0.125	Fuel Oil #2	BP, 1996
Toluene	7	7.58	2.73 - 21.80	Gasoline	LUFT, 1988
			0.09 - 2.5	Crude Oil	API, 1993
			1.33	JP-4	API, 1993
			0.007 - 0.70	Diesel	BP, 1996
			0.025 - 0.110	Fuel Oil #2	BP, 1996
Ethylbenzene	8	8.5	0.36 - 2.86	Gasoline	LUFT, 1988
			0.09 - 0.31	Crude Oil	API, 1993
			0.37	JP-4	API, 1993
			0.007 - 0.20	Diesel	BP, 1996
			0.028 - 0.04	Fuel Oil #2	BP, 1996
o-Xylene	8	8.81	0.68 - 2.86	Gasoline	LUFT, 1988
			0.03 - 0.68	Crude Oil	API, 1993
			1.01	JP-4	API, 1993
			0.09	JP-5	API, 1993
			0.06	JP-8	API, 1993
			.001 - 0.085	Diesel	BP, 1996
m-Xylene	8	8.6	1.77 - 3.87	Gasoline	LUFT, 1988
			0.08 - 2.0	Crude Oil	API, 1993
			0.96	JP-4	API, 1993
			0.13	JP-5	API, 1993
			0.06	JP-8	API, 1993
			0.018 - 0.512	Diesel	BP, 1996

Table 2. *Continued*

Compound	Number of Carbons	EC	Weight Percent	Fuel Type	Reference
p-Xylene	8	8.61	0.77 - 1.58	Gasoline	LUFT, 1988
			0.09 - 0.68	Crude Oil	API, 1993
			0.35	JP-4	API, 1993
			0.018 - 0.512	Diesel	BP, 1996
Styrene	9	8.83	<.002	Diesel	BP, 1996
1-Methyl-4-ethylbenzene	9	9.57	0.18 - 1.00	Gasoline	LUFT, 1988
			0.03 - 0.13	Crude Oil	API, 1993
			0.43	JP-4	API, 1993
1-Methyl-2-ethylbenzene	9	9.71	0.19 - 0.56	Gasoline	LUFT, 1988
			0.01 - 0.09	Crude Oil	API, 1993
			0.23	JP-4	API, 1993
1-Methyl-3-ethylbenzene	9	9.55	0.31 - 2.86	Gasoline	LUFT, 1988
			0.04 - 0.4	Crude Oil	API, 1993
			0.49	JP-4	API, 1993
1-Methyl-2-n-propylbenzene	10		0.01 - 0.17	Gasoline	LUFT, 1988
1-Methyl-3-n-propylbenzene	10		0.08 - 0.56	Gasoline	LUFT, 1988
1-Methyl-2-isopropylbenzene	10		0.01 - 0.12	Gasoline	LUFT, 1988
			0.29	JP-4	API, 1993
			0.56	JP-8	API, 1993
1-Methyl-3-isopropylbenzene	10	10.09			
1-Methyl-4-isopropylbenzene	10	10.13	0.003 - 0.026	Diesel	BP, 1996
1-Methyl-3-t-butylbenzene	11		0.03 - 0.11	Gasoline	LUFT, 1988
1-Methyl-4-t-butylbenzene	11	10.92	0.04 - 0.13	Gasoline	LUFT, 1988
1,2-Dimethyl-3-ethylbenzene	10	10.93	0.02 - 0.19	Gasoline	LUFT, 1988
1,2-Dimethyl-4-ethylbenzene	10	10.75	0.50 - 0.73	Gasoline	LUFT, 1988
			0.77	JP-4	API, 1993
1,3-Dimethyl-2-ethylbenzene	10	10.81	0.21 - 0.59	Gasoline	LUFT, 1988
1,3-Dimethyl-4-ethylbenzene	10	10.75	0.03 - 0.44	Gasoline	LUFT, 1988
1,3-Dimethyl-5-ethylbenzene	10	10.51	0.11 - 0.42	Gasoline	LUFT, 1988
			0.61	JP-4	API, 1993
			0.62	JP-8	API, 1993

Table 2. *Continued*

Compound	Number of Carbons	EC	Weight Percent	Fuel Type	Reference
1,3-Dimethyl-5-t-butylbenzene	12		0.02 - 0.16	Gasoline	LUFT, 1988
1,4-Dimethyl-2-ethylbenzene	10	10.68	0.05 - 0.36	Gasoline	LUFT, 1988
			0.7	JP-4	API, 1993
1,2,3-Trimethylbenzene	9	10.06	0.21 - 0.48	Gasoline	LUFT, 1988
			0.1	Crude Oil	API, 1993
1,2,4-Trimethylbenzene	9	9.84	0.66 - 3.30	Gasoline	LUFT, 1988
			0.13 - 0.69	Crude Oil	API, 1993
			1.01	JP-4	API, 1993
			0.37	JP-5	API, 1993
			0.27	JP-8	API, 1993
1,3,5-Trimethylbenzene	9	9.62	0.13 - 1.15	Gasoline	LUFT, 1988
			0.05 - 0.18	Crude Oil	API, 1993
			0.42	JP-4	API, 1993
			0.09 - 0.24	Diesel	BP, 1996
1,2,3,4-Tetramethylbenzene	10	11.57	0.02 - 0.19	Gasoline	LUFT, 1988
			0.2	Crude Oil	API, 1993
1,2,3,5-Tetramethylbenzene	10	11.09	0.14 - 1.06	Gasoline	LUFT, 1988
1,2,4,5-Tetramethylbenzene	10	11.05	0.05 - 0.67	Gasoline	LUFT, 1988
1,2-Diethylbenzene	10	10.52	0.57	Gasoline	LUFT, 1988
1,3-Diethylbenzene	10	10.4	0.05 - 0.38	Gasoline	LUFT, 1988
			0.46	JP-4	API, 1993
			0.61	JP-5	API, 1993
1,4-Diethylbenzene	10	10.46	0.77	JP-5	API, 1993
1,2,4-Triethylbenzene	12	12.29	0.72	JP-5	API, 1993
			0.99	JP-8	API, 1993
1,3,5-Triethylbenzene	12	12.1	0.6	JP-8	API, 1993
n-Propylbenzene	9	9.47	0.08 - 0.72	Gasoline	LUFT, 1988
			0.71	JP-4	API, 1993
			0.03 - 0.048	Diesel	BP, 1996
Isopropylbenzene	9	9.13	<10.01 - 0.23	Gasoline	LUFT, 1988
			0.3	JP-4	API, 1993
			<0.01	Diesel	BP, 1996

Table 2. *Continued*

Compound	Number of Carbons	EC	Weight Percent	Fuel Type	Reference
n-Butylbenzene	10	10.5	0.04 - 0.44	Gasoline	LUFT, 1988
			0.031 - 0.046	Diesel	BP, 1996
Isobutylbenzene	10	9.96	0.01 - 0.08	Gasoline	LUFT, 1988
sec-Butylbenzene	10	9.98	0.01 - 0.13	Gasoline	LUFT, 1988
t-Butylbenzene	10	9.84	0.12	Gasoline	LUFT, 1988
1-t-Butyl-3,4,5-trimethylbenzene	13		0.24	JP-5	API, 1993
n-Pentylbenzene	11	11.49	0.01 - 0.14	Gasoline	LUFT, 1988
Isopentylbenzene	11		0.07 - 0.17	Gasoline	LUFT, 1988
n-Hexylbenzene	12	12.5			
n-Heptylbenzene	13		0.27	JP-5	API, 1993
			0.25	JP-8	API, 1993
n-Octylbenzene	14		0.78	JP-5	API, 1993
			0.61	JP-8	API, 1993
Biphenyl	12	14.26	0.006 - .04	Crude Oil	API, 1993
			0.7	JP-5	API, 1993
			0.63	JP-8	API, 1993
			0.01 - 0.12	Diesel	BP, 1996
			0.006 - 0.009	Fuel Oil #2	BP, 1996
4-Methylbiphenyl	13	14.92			
4,4'-Dimethylbiphenyl	14	16.55			
Phenylcyclohexane	12		0.82	JP-5	API, 1993
			0.87	JP-8	API, 1993
NAPHTHENO BENZENES					
Acenaphthene	12	15.5	0.013 - 0.022	Fuel Oil #2	BP, 1996
Acenaphthylene	12	15.00	0.006	Fuel Oil #2	BP, 1996
Indan	9	10.27	0.25 - 0.34	Gasoline	LUFT, 1988
			0.07	Crude Oil	API, 1993
1-Methylindan	10		0.04 - 0.17	Gasoline	LUFT, 1988
2-Methylindan	10	11.39	0.02 - 0.10	Gasoline	LUFT, 1988

Table 2. *Continued*

Compound	Number of Carbons	EC	Weight Percent	Fuel Type	Reference
4-Methylindan	10	11.33	0.01 - 0.16	Gasoline	LUFT, 1988
5-Methylindan	10	11.28	0.09 - 0.30	Gasoline	LUFT, 1988
Tetralin (tetrahydronaphthalene)	10	11.7	0.01 - 0.14	Gasoline	LUFT, 1988
			0.03	Crude Oil	API, 1993
5-Methyltetrahydronaphthalene	11		0.08	Crude Oil	API, 1993
6-Methyltetrahydronaphthalene	11		0.09	Crude Oil	API, 1993
Fluorene	13	16.55	0.003 - 0.06	Crude Oil	API, 1993
			0.034 - 0.15	Diesel	BP, 1996
			0.004 - 0.045	Fuel Oil #2	BP, 1996
1-Methylfluorene	14	17.99			
Fluoranthene	16	21.85	0.0000007 - 0.02	Diesel	BP, 1996
			0.000047 - 0.00037	Fuel Oil #2	BP, 1996
2,3- Benzofluorene	17	23.83	<0.0024	Fuel Oil #2	BP, 1996
1,2- Benzofluorene	17	24.2			
Benzo(a)fluorene	17		<0.0006	Fuel Oil #2	BP, 1996
Benzo(ghi)fluoranthene	18		<0.0024	Fuel Oil #2	BP, 1996
Benz(b)fluoranthene	20	30.14	0.0000003 - 0.000194	Diesel	BP, 1996
			<0.0024	Fuel Oil #2	BP, 1996
Benz(k)fluoranthene	20	30.14	0.0000003 - 0.000195	Diesel	BP, 1996
			<0.00006	Fuel Oil #2	BP, 1996
Indeno (1,2,3-cd) pyrene	22	35.01	0.000001 - 0.000097	Diesel	BP, 1996
			<0.0012	Fuel Oil #2	BP, 1996
ALKYL NAPHTHALENES					
Naphthalene	10	11.69	0.09 - 0.49	Gasoline	LUFT, 1988
			0.02 - 0.09	Crude Oil	API, 1993
			0.5	JP-4	API, 1993
			0.57	JP-5	API, 1993
			1.14	JP-8	API, 1993
			0.01 - 0.80	Diesel	BP, 1996
			0.009 - 0.40	Fuel Oil #2	BP, 1996

Table 2. *Continued*

Compound	Number of Carbons	EC	Weight Percent	Fuel Type	Reference
1-Methylnaphthalene	11	12.99	0.78	JP-4	API, 1993
			1.44	JP-5	API, 1993
			1.84	JP-8	API, 1993
			0.001 - 0.81	Diesel	BP, 1996
			0.29 - 0.48	Fuel Oil #2	BP, 1996
2-Methylnaphthalene	11	12.84	0.56	JP-4	API, 1993
			1.38	JP-5	API, 1993
			1.46	JP-8	API, 1993
			0.001 - 1.49	Diesel	BP, 1996
			0.36 -1.00	Fuel Oil #2	BP, 1996
1,3-Dimethylnaphthalene	12	14.77	0.55 - 1.28	Diesel	BP, 1996
1,4-Dimethylnaphthalene	12	14.6	0.110 - 0.23	Diesel	BP, 1996
			0.043 - 0.045	Fuel Oil #2	BP, 1996
1,5-Dimethylnaphthalene	12	13.87	0.16 - 0.36	Diesel	BP, 1996
2,3-Dimethylnaphthalene	12	15	0.46	JP-5	API, 1993
			0.36	JP-8	API, 1993
2,6-Dimethylnaphthalene	12	14.6	0.25	JP-4	API, 1993
			1.12	JP-5	API, 1993
			1.34	JP-8	API, 1993
1-Ethylnaphthalene	12	14.41	0.32	JP-5	API, 1993
			0.33	JP-8	API, 1993
2-Ethylnaphthalene	12	13.99			
1,4,5-Trimethylnaphthalene	13	10.6			
1-Phenylnaphthalene	16				
POLYNUCLEAR AROMATICS					
Anthracene	14	19.43	0.000003 -0.02	Diesel	BP, 1996
			0.00010 - 0.011	Fuel Oil #2	BP, 1996
2-Methyl anthracene	15	20.73	0.000015 - 0.018	Diesel	BP, 1996
			0.009 - 0.017	Fuel Oil #2	BP, 1996
9-Methyl anthracene	15	20.45			
2-Ethyl anthracene	16				
9,10-Dimethyl anthracene	16		0.002 - 0.006	Fuel Oil #2	BP, 1996

Table 2. *Continued*

Compound	Number of Carbons	EC	Weight Percent	Fuel Type	Reference
Phenanthrene	14	19.36	0.003 - 0.05	Crude Oil	API, 1993
			0.000027 - 0.30	Diesel	BP, 1996
			0.009 -0.170	Fuel Oil #2	BP, 1996
1-Methylphenanthrene	15	20.73	0.000011 - 0.024	Diesel	BP, 1996
			0.017	Fuel Oil #2	BP, 1996
2-Methylphenanthrene	15		0.014 - 0.18	Diesel	BP, 1996
			0.768	Fuel Oil #2	BP, 1996
3-Methylphenanthrene	15		0.000013 - 0.011	Diesel	BP, 1996
4 & 9-Methylphenanthrene	15		0.00001 - 0.034	Diesel	BP, 1996
Pyrene	16	20.8	Not quantified	Gasoline	LUFT, 1988
			0.000018 - 0.015	Diesel	BP, 1996
			0.00 - 0.012	Fuel Oil #2	BP, 1996
1-Methylpyrene	17		0.0000024 - 0.00137	Diesel	BP, 1996
2-Methylpyrene	17		0.0000037 - 0.00106	Diesel	BP, 1996
Benz(a)anthracene	18	26.37	Not quantified	Gasoline	LUFT, 1988
			0.0000021 - 0.00067	Diesel	BP, 1996
			0.000002 - 0.00012	Fuel Oil #2	BP, 1996
Chrysene	18	27.41	0.000045	Diesel	BP, 1996
			0.000037 - 0.00039	Fuel Oil #2	BP, 1996
Triphenylene	18	26.61	0.00033	Diesel	BP, 1996
			0.00002 - 0.00014	Fuel Oil #2	BP, 1996
Cyclopenta(cd)pyrene	18		0.000002 - 0.0000365	Diesel	BP, 1996
1-Methyl-7-isopropylphenanthrene	18		0.0000015 - 0.00399	Diesel	BP, 1996
3-Methylchrysene	19		<0.001	Diesel	BP, 1996
5-Methylchrysene	19				
6-Methylchrysene	19		<0.0005	Diesel	BP, 1996
Benzo(b)chrysene			<0.0036	Fuel Oil #2	BP, 1996
Benz(a)pyrene	20	31.34	0.19 - 2.8 mg/kg	Gasoline	LUFT, 1988
			0.000005 -0.00084	Diesel	BP, 1996
			0.000001 - 0.000060	Fuel Oil #2	BP, 1996

Table 2. *Continued*

Compound	Number of Carbons	EC	Weight Percent	Fuel Type	Reference
Benz(e)pyrene	20	31.17	Not quantified	Gasoline	LUFT, 1988
			0.0000054 - 0.000240	Diesel	BP, 1996
			0.0000020 - 0.000010	Fuel Oil #2	BP, 1996
Benzo(ghi)pyrene			0.0000010 - 0.0000070	Fuel Oil #2	BP, 1996
Perylene	20	31.34	<0.0001	Diesel	BP, 1996
			<0.0024	Fuel Oil #2	BP, 1996
3-Methylcholanthrene	21		<0.00006	Fuel Oil #2	BP, 1996
Benz(ghi)perylene	22	34.01	Not quantified	Gasoline	LUFT, 1988
			0.0000009 - 0.00004	Diesel	BP, 1996
			0.0000057	Fuel Oil #2	BP, 1996
Picene	22		0.0000004 - 0.000083	Diesel	BP, 1996
			<0.00012	Fuel Oil #2	BP, 1996
1,2,5,6-Dibenz anthracene	22	33.92			
Coronene	24	34.01	<0.000024	Fuel Oil #2	BP, 1996

For each compound in Table 2 an attempt was made to find values for each of the properties listed in Table 1 (Table 3). This work was done by EA Engineering, Inc. and the Analytical Technical Advisory Committee. References for values listed in Table 3 are included in Appendix B.

3.2.1 Data Sources for Physical and Chemical Properties

Values for all relevant compound-specific parameters were either collected or derived from the references searched. Comprehensive data for many of the compounds are conveniently compiled in the *Illustrated Handbook of Physical-Chemical Properties and Environmental Fate for Organic Chemicals* (Mackay et al., 1993). Appendix A presents a reformatted Table 2 with chemical constituents listed by fuel type.

In many cases, a large number of values were listed for the same parameters at similar temperatures, and in some cases the values spanned a range of more than an order of magnitude. For example, more than 20 values were listed for aqueous solubility of n-octane, ranging from 0.43 to 14 mg/L (these include both experimental values and values either estimated or calculated from other known properties, such as activity coefficients). Differences in reported values can be attributed to many factors, including different measuring techniques (with their inherent errors in precision and accuracy), experimental conditions, the objectives of experimental measurements, and the assumptions used in making estimates. At constant temperature and pressure, a given chemical solubility in water should not vary by any significant amount.

Mackay et al. (1993) evaluated the published values of water solubility, vapor pressure, Henry's law constant, and octanol-water partition coefficients for many of the compounds considered in this report and selected a "best" or "most likely" value. For compounds or parameters for which a "best" or "most likely" value was not provided, other generally accepted references were used. Overall, attempts were made to use actual measured values instead of estimated values when available.

When possible, parameters were reported for standard atmospheric pressure at temperatures between 10°C and 25°C. The references for values reported in Table 3 may be found in Appendix B.

3.2.2 Estimation Methods

Because of a lack of data or high variability in the values reported in the scientific literature for critical parameters, primarily K_{ow}, K_{oc}, and D_{air}, some of these were estimated, using the techniques described below. Estimated values are set in boldface in Table 3. Bold values represent only those values the Working Group has estimated, not estimated values presented in the literature. Overall, most of the values are measured, establishing a real database of measured values for use in this and future applications.

3.2.2.1 Octanol-Water Partition Coefficient, K_{ow}

The base 10 logarithm of K_{ow} was estimated using the commercially available software program, ClogP (Leo and Hansch, 1991). ClogP uses a fragment constant

Table 3. Physical-Chemical Properties of Petroleum Hydrocarbons

Compound	Carbons	EC	MW	S(mg/L)	SG	VP(atm)	H(c/c)	log K_{ow}	K_{oc}	BP(C)	D air(cm^2/s)	Dw(cm^2/s)	CAS #	LF	VF
ALKYL BENZENES															
Benzene	6	6.50	78.11	1780	0.8765	1.25E-01	2.25E-01	2.13	**8.12E+01**	80.1	8.80E-02	9.80E-06	71-43-2	9.00E-02	5.61E-04
Toluene	7	7.58	92.13	515	0.8669	3.75E-02	2.74E-01	2.69	**2.34E+02**	110.6	8.70E-02	8.60E-06	108-88-3	3.37E-02	2.53E-04
Ethylbenzene	8	8.50	106.2	152	0.867	1.25E-02	3.58E-01	3.13	**5.37E+02**	136.2	7.50E-02	7.80E-06	100-41-4	1.50E-02	1.27E-04
m-Xylene	8	8.60	106.2	160	0.8842	1.09E-02	2.95E-01	3.20	**6.12E+02**	139	7.00E-02	7.80E-06	108-38-3	1.32E-02	8.60E-05
p-Xylene	8	8.61	106.2	215	0.8611	1.15E-02	2.33E-01	3.18	**5.90E+02**	138			106-42-3	1.38E-02	
o-Xylene	8	8.81	106.2	220	0.8802	1.15E-02	2.28E-01	3.15	**5.57E+02**	144	8.70E-02	1.00E-05	95-47-6	1.45E-02	9.09E-05
Styrene	8	8.83	104.14	300	0.906	7.90E-03	1.23E-01	3.05	**4.61E+02**	145.2	7.10E-02	8.00E-06	100-42-5	1.76E-02	4.84E-05
Isopropylbenzene	9	9.13	120.2	50	0.8618	6.02E-03	5.92E-01	3.63	**1.38E+03**	154.2	6.50E-02	7.10E-06	98-82-8	5.92E-03	7.17E-05
n-Propylbenzene	9	9.47	120.2	52	0.862	4.44E-03	4.20E-01	3.69	**1.54E+03**	159.2	**5.98E-02**	7.83E-06	103-65-1	5.30E-03	4.19E-05
1-Methyl-3-ethylbenzene	9	9.55	120.2		0.8645	3.86E-03		**3.63**	**1.38E+03**	161.5	**5.65E-02**		620-14-4		
1-Methyl-4-ethylbenzene	9	9.57	120.2	95	0.8614	3.90E-03	2.02E-01	3.63	**1.38E+03**	162	**5.76E-02**		622-96-8	5.94E-03	
1,3,5-Trimethylbenzene	9	9.62	120.2	50	0.8652	3.21E-03	3.15E-01	3.58	**1.25E+03**	164.7	**6.28E-02**	7.85E-06	108-67-8	6.52E-03	4.06E-05
1-Methyl-2-ethylbenzene	9	9.71	120.2	75	0.8867	3.26E-03	2.14E-01	3.63	**1.38E+03**	165.2	**5.83E-02**		611-14-3	5.94E-03	
1,2,4-Trimethylbenzene	9	9.84	120.2	57	0.8758	2.66E-03	2.30E-01	3.60	**1.30E+03**	169.4	**6.04E-02**	7.85E-06	95-63-6	6.29E-03	2.75E-05
t-Butylbenzene	10	9.84	134.22	30	0.8665	2.82E-03	5.17E-01	4.11	**3.41E+03**	169	**6.27E-02**	7.29E-06	98-06-6	2.41E-03	2.46E-05
Isobutylbenzene	10	9.96	134.22	10.1	0.8532	2.47E-03	1.34E+00	4.01	**2.83E+03**	170	**5.98E-02**		538-93-2	2.89E-03	
sec-Butylbenzene	10	9.98	134.22	17	0.8621	2.37E-03	7.63E-01	**4.10**	**2.83E+03**	173	**6.27E-02**	7.29E-06	135-98-8	2.90E-03	4.37E-05
1,2,3-Trimethylbenzene	9	10.05	120.2	70	0.8944	1.97E-03	1.38E-01	3.55	**1.19E+03**	176.1	**6.02E-02**		526-73-8	6.91E-03	
1-Methyl-3-isopropylbenzene	10	10.09						4.10	**3.35E+03**						
1-Methyl-4-isopropylbenzene	10	10.13	134.22	34	0.857	2.01E-03	3.25E-01	4.10	**3.35E+03**	177.1			99-87-6	2.46E-03	
1,3-Diethylbenzene	10	10.40						**4.10**	**3.35E+03**						
1,4-Diethylbenzene	10	10.46						**4.10**	**3.35E+03**		**5.94E-02**				

Note: Bold values are estimated

Table 3. *Continued*

Compound	Carbons	EC	MW	S(mg/L)	SG	VP(atm)	H(c/c)	log K_{ow}	K_{oc}	BP(C)	D air(cm^2/s)	Dw(cm^2/s)	CAS #	LF	VF
n-Butylbenzene	10	10.50	134.22	13.8	0.8601	1.35E-03	5.38E-01	4.26	**4.53E+03**	183	**5.98E-02**		104-51-8	1.82E-03	
1,3-Dimethyl-5-ethylbenzene	10	10.51						**4.14**							
1,2-Diethylbenzene	10	10.52						**4.10**	**3.35E+03**						
1,4-Dimethyl-2-ethylbenzene	10	10.68						**4.14**	**3.61E+03**						
1,2-Dimethyl-4-ethylbenzene	10	10.75						**4.14**	**3.61E+03**						
1,3-Dimethyl-2-ethylbenzene	10	10.81						**4.14**	**3.61E+03**						
1-Methyl-4-t-butylbenzene	11	10.92						**4.73**	**1.10E+04**						
1,2-Dimethyl-3-ethylbenzene	10	10.93						**4.14**	**3.61E+03**						
1,2,4,5-Tetramethylbenzene	10	11.05	134.22	3.48	0.838	6.51E-04	1.03E+00	4.10	**3.35E+03**	196.8	**5.61E-02**		95-93-2	2.45E-03	
1,2,3,5-Tetramethylbenzene	10	11.09	134.22		0.8585	6.12E-04		4.04	**2.99E+03**	198			527-53-7		
n-Pentylbenzene	11	11.49	148.25	3.85	0.8585	4.34E-04	6.84E-01	4.90	**1.52E+04**	205.4			538-68-1	5.44E-04	
1,2,3,4-Tetramethylbenzene	10	11.57	134.22		0.9052	4.44E-04		3.90	**2.30E+03**	205			48-23-3		
1,3,5-Triethylbenzene	12	12.10						**5.23**	**2.83E+04**						
1,2,4-Triethylbenzene	12	12.29						**5.23**	**2.83E+04**						
n-Hexylbenzene	12	12.50	162.28	1.02	0.861	1.34E-04	8.74E-01	5.52	**4.89E+04**	226			1077-16-3	1.69E-04	
4-Methylbiphenyl	13	14.92	168.24	4.05	1.105			4.63	**9.11E+03**	267.5			644-08-6		
4,4'-Dimethylbiphenyl	14	16.55	182.27	0.175	0.917			5.09	**2.17E+04**	295			613-33-2		
Isopentylbenzene	11							**4.90**	**1.52E+04**						
1,3-Dimethyl-5-t-butylbenzene	12							**5.23**	**2.83E+04**						
Phenylcyclohexane	12							4.99	**1.80E+04**						
1-t-Butyl-3,4,5-trimethylbenzene	13							**5.72**	**7.14E+04**						
n-Heptylbenzene	13							**5.82**	**8.62E+04**						
n-Octylbenzene	14							**6.34**	**2.30E+05**						

Table 3. *Continued*

Compound	Carbons	EC	MW	S(mg/L)	SG	VP(atm)	H(c/c)	log K_{ow}	K_{oc}	BP(C)	D air(cm^2/s)	Dw(cm^2/s)	CAS #	LF	VF
Xylene (mixed isomers)	8		106.2	180		5.76E+00			**2.40E+02**		7.22E-02	8.50E-06	1330207		
1-Methyl-2-n-propylbenzene	10							**4.10**	**3.35E+03**						
1-Methyl-3-n-propylbenzene	10							**4.10**	**3.35E+03**						
1-Methyl-2-isopropylbenzene	10							**4.10**	**3.35E+03**						
1,3-Dimethyl-4-ethylbenzene	10							**4.14**	**3.61E+03**						
1-Methyl-3-t-butylbenzene	11							**4.73**	**1.10E+04**						
ALKYL NAPHTHALENES															
1,4,5-Trimethylnaphthalene	13	10.60	176.2	2.1		1.64E-05	2.31E-02	5.00	**1.83E+04**	185			2131-41-1	4.50E-04	
Naphthalene	10	11.69	128.19	31	1.03	3.63E-04	1.74E-02	3.37	**8.44E+02**	218	5.90E-02	7.50E-06	91-20-3	9.70E-03	3.13E-06
2-Methylnaphthalene	11	12.84	142.2	25	1.0058	1.11E-04	2.07E-02	3.86	**2.13E+03**	241.9	**5.60E-02**	7.84E-06	91-57-6	3.86E-03	1.41E-06
1-Methylnaphthalene	11	12.99	142.2	28	1.022	8.72E-05	1.81E-02	3.87	**2.17E+03**	244.6	**5.70E-02**		90-12-0	3.79E-03	
1,5-Dimethylnaphthalene	12	13.87	156.23	3.1				4.38	**5.68E+03**	249			571-61-9		
2-Ethylnaphthalene	12	13.99	156.23	8	0.992	3.95E-05	3.15E-02	**4.40**	**5.90E+03**	251.2			939-27-5	1.40E-03	
1-Ethylnaphthalene	12	14.12	156.23	10.1	1.0082	2.48E-05	1.57E-02	4.40	**5.90E+03**	258.7			1127-76-0	1.40E-03	
2,6-Dimethylnaphthalene	12	14.30	156.23	1.7	1.142	9.15E-05	5.19E-02	4.31	**4.98E+03**	262			581-40-2	1.66E-03	
1,4-Dimethylnaphthalene	12	14.60	156.23	11.4	1.0166	2.24E-05	1.26E-02	4.37	**5.58E+03**	262			571-58-4	1.48E-03	
1,3-Dimethylnaphthalene	12	14.77	156.23	8	1.0144			4.42	**6.13E+03**	265			575-41-7		
2,3-Dimethylnaphthalene	12	15.00	156.23	2.5	1.003	6.09E-05	2.52E-02	4.40	**5.90E+03**	269			581-40-8	1.40E-03	
1-Phenylnaphthalene	16							**5.06**	**2.05E+04**						

Table 3. *Continued*

Compound	Carbons	EC	MW	S(mg/L)	SG	VP(atm)	H(c/c)	log K_{ow}	K_{oc}	BP(C)	D air(cm^2/s)	Dw(cm^2/s)	CAS #	LF	VF
NAPHTHENO BENZENES															
Indan	9	10.27	118.18	100	0.9639	1.94E-03	9.40E-02	3.33	**7.83E+02**	178			496-11-7	1.04E-02	
5-Methylindan	11	11.28						**3.65**	**1.43E+03**						
4-Methylindan	11	11.33						**3.65**	**1.43E+03**						
2-Methylindan	11	11.39						**3.65**	**1.43E+03**						
Tetralin (tetrahydronaphthalene)	10	11.70	132.21	15	0.9695	5.23E-04	1.89E-01	3.83	**2.01E+03**	207.65			119-64-2	4.08E-03	
Biphenyl	12	14.26	154.2	7	1.04	3.66E-05	1.16E-02	3.90	**2.30E+03**	256			92-52-4	3.58E-03	
Acenaphthylene	12	15.06	152.2	16.1	0.89	4.09E-05	3.39E-03	4.00	**2.77E+03**	270	4.40E-02	7.53E-06	208-96-8	2.97E-03	1.40E-07
Acenaphthene	12	15.50	154.21	3.8	1.19	1.50E-05	4.91E-03	3.92	**2.38E+03**	277.5	4.21E-02	7.69E-06	83-32-9	3.45E-03	2.26E-07
Fluorene	13	16.55	166.2	1.9	1.202	7.06E-06	3.19E-03	4.18	**3.90E+03**	295	3.60E-02	7.88E-06	86-73-7	2.12E-03	7.69E-08
1-Methylfluorene	14	17.99	180.25	1.09				4.97	**1.73E+04**	318			1730-37-6		
Fluoranthene	16	21.85	202.3	0.26	1.252	8.61E-08	4.17E-04	5.22	**2.78E+04**	375	3.02E-02	6.35E-06	206-44-0	2.97E-04	1.23E-09
2,3- Benzofluorene	17	23.83	216.3	0.002				5.75	**7.55E+04**	402			243-17-4		
1,2- Benzofluorene	17	24.20	216.3	0.0454				5.40	**3.90E+04**	407			238-84-6		
Benz(b)fluoranthene	20	30.14	252.32	0.0015		6.67E-08		5.80	**8.30E+04**	481	2.26E-02	5.56E-06	205-99-2		
Benz(k)fluoranthene	20	30.14	252.32	0.0008		4.07E-11	6.46E-06	6.00	**1.21E+05**	481	2.26E-02	5.56E-06	207-08-9	6.82E-05	1.22E-11
Indeno (1,2,3-cd) pyrene	21	35.01	276.3	0.062		1.00E-09	2.07E-11	7.00	**8.00E+05**	536	2.30E-02	4.41E-06	53-70-3	1.03E-05	7.69E-13
5-Methyltetrahydronaphthalene	11							**4.40**	**5.90E+03**						
6-Methyltetrahydronaphthalene	11							**4.40**	**5.90E+03**						
1-Methylindan	11							**3.65**	**1.43E+03**						

Table 3. *Continued*

Compound	Carbons	EC	MW	S(mg/L)	SG	VP(atm)	H(c/c)	log K_{ow}	K_{oc}	BP(C)	D air(cm^2/s)	Dw(cm^2/s)	CAS #	LF	VF
POLYNUCLEAR AROMATICS															
Phenanthrene	14	19.36	178.2	1.1	1.03	1.12E-06	1.31E-03	4.57	**8.14E+03**	339	3.30E-02	7.47E-06	85-01-8	1.01E-03	1.40E-08
Anthracene	14	19.43	178.2	0.045	1.283	7.68E-07	1.60E-03	4.54	**7.69E+03**	340	3.24E-02	7.74E-06	120-12-7	1.07E-03	1.77E-08
9-Methyl anthracene	15	20.45	192.26	0.261	1.065		4.30E-02	5.07	**2.09E+04**	355			779-02-2	3.95E-04	
2-Methyl anthracene	15	20.73	192.26	0.03	1.81			5.15	**2.43E+04**	359			613-12-7		
1-Methylphenanthrene	15	20.73	192.26	0.27				5.14	**2.39E+04**	359			832-69-6		
Pyrene	16	20.30	202.3	0.132	1.271	1.17E-07	3.71E-04	5.18	**2.57E−04**	360	2.70E-02	7.24E-06	129-00-0	3.21E-04	1.07E-09
Benz(a)anthracene	18	26.37	228.3	0.011	1.2544	5.98E-09	2.34E-04	5.91	**1.02E−05**	435	5.10E-02	9.00E-06	56-55-3	8.08E-05	3.22E-10
Triphenylene	18	26.61	228.3	0.043	1.302	1.19E-09	4.84E-06	5.49	**4.62E−04**	438			217-59-4	1.79E-04	
Chrysene	18	27.41	228.3	0.0015	1.274	1.06E-09	1.80E-04	5.79	**8.14E−04**	448	2.48E-02	6.21E-06	218-01-9	1.01E-04	1.58E-10
Benz(e)pyrene	20	31.17	252.3	0.004		2.38E-10	8.07E-06	6.44	**2.78E+05**	493			192-97-2	2.97E-05	
Benz(a)pyrene	20	31.34	252.3	0.0038		2.10E-10	1.86E-05	6.04	**1.31E+05**	495	4.30E-02	9.00E-06	50-32-8	6.32E-05	2.96E-11
Perylene	20	31.34	252.32	0.0004	1.35		1.21E-06	6.25	**1.94E+05**	495			198-55-0	4.25E-05	
1,2,5,6-Dibenz anthracene	22	33.92	278.4	0.0005	1.28	1.33E-08	3.07E-06	6.75	**4.99E+05**	524	2.00E-02	5.24E-06	53-70-3	1.66E-05	2.40E-12
Benz(ghi)perylene	21	34.01	268.36	0.0003		2.22E-10	3.03E-05	6.50	**3.11E+05**	525	4.90E-02	5.56E-06	191-24-2	2.65E-05	1.59E-11
Coronene	24	34.01	300.36	0.0001				6.75	**4.99E+05**	525			191-07-1		
3-Methylcholanthrene	21		268.36	0.0019	1.28	3.41E-10	5.85E-05	6.42	**2.68E+05**				56-49-5	3.09E-05	
Picene	22		278.3	0.0004				7.19	**1.14E+06**						
2-Ethyl anthracene	16		206.3	0.03				5.15	**2.43E+04**						
9,10-Dimethyl anthracene	16		206.3	0.056				5.25	**2.94E+04**				781-43-1		
1-Methylpyrene	17							**5.45**	**4.29E+04**						
5-Methylchrysene	19		242.1	0.0073				6.42	**2.68E+05**						

Table 3. *Continued*

Compound	Carbons	EC	MW	S(mg/L)	SG	VP(atm)	H(c/c)	log K_{ow}	K_{oc}	BP(C)	D air(cm^2/s)	Dw(cm^2/s)	CAS #	LF	VF
STRAIGHT CHAIN ALKANES															
Propane	3	3.00	44.09	62.4	0.58	8.50E+00	2.89E+01	2.36	**1.25E+02**	-42	1.09E-01		74-98-6	1.44E-02	
n-Butane	4	4.00	58.13	61.4	0.5786	2.40E+00	3.87E+01	2.89	**3.41E+02**	-0.5	9.57E-02		106-97-8	8.78E-03	
n-Pentane	5	5.00	72.15	38.5	0.6262	6.75E-01	5.17E+01	3.45	**9.82E+02**	36.07	8.17E-02		109-66-0	4.64E-03	
n-Hexane	6	6.00	86.17	9.5	0.6593	1.99E-01	7.39E+01	4.11	**3.41E+03**	68.95	2.00E-01	**7.77E-06**	110-54-3	1.81E-03	8.45E-03
n-Heptane	7	7.00	100.21	2.93	0.6837	6.03E-02	8.43E+01	5.00	**1.83E+04**	98.42	6.59E-02	**7.59E-06**	142-82-5	4.21E-04	7.37E-04
n-Octane	8	8.00	114.23	0.66	0.7027	1.78E-02	1.26E+02	5.15	**2.43E+04**	125.7	5.98E-02	**7.13E-06**	111-65-9	3.15E-04	7.45E-04
n-Nonane	9	9.00	128.26	0.22	0.7177	5.64E-03	1.34E+02	5.65	**6.25E+04**	150.8	**5.30E-02**	**5.97E-06**	111-84-2	1.28E-04	2.87E-04
n-Decane	10	10.00	142.29	0.052	0.7301	1.73E-03	1.93E+02	6.25	**1.94E+05**	174.1	**5.00E-02**	**5.62E-06**	124-18-5	4.19E-05	1.28E-04
n-Undecane	11	11.00	156.32	0.04	0.7402	5.15E-04	7.49E+01	6.94	**7.14E+05**	195.9	**4.70E-02**	**5.31E-06**	1120-21-4	1.15E-05	1.28E-05
n-Dodecane	12	12.00	170.33	0.0037	0.7487	1.55E-04	3.17E+02	7.24	**1.26E+06**	216.3	**4.50E-02**	**5.06E-06**	112-40-3	6.54E-06	2.94E-05
n-Tridecane	13	13.00	185.36		0.755	9.54E-05		**7.57**	**2.35E+06**		**4.20E-02**	**4.78E-06**			
n-Tetradecane	14	14.00	198.4	0.0007		3.83E-05	1.56E+02	7.20	**1.17E+06**	252	**4.00E-02**	**4.60E-06**		7.06E-06	1.39E-05
n-Pentadecane	15	15.00	212.42		0.769	1.53E-05		**8.63**	**1.74E+07**	270	**3.90E-02**	**4.43E-06**			
n-Hexadecane	16	16.00	226.4	5E-05	0.7749	6.30E-06	1.57E+02	8.25	**8.47E+06**	287	**3.70E-02**	**4.20E-06**	544763	9.74E-07	1.78E-06
n-Heptadecane	17	17.00	240.4		0.778	2.68E-06		**9.69**	**1.28E+08**	303	**3.60E-02**	**4.09E-06**			
n-Octadecane	18	18.00	254.4	4E-06	0.777	1.14E-06	2.51E+02	9.32	**6.39E+07**	317	**3.30E-02**	**3.75E-06**		1.29E-07	3.37E-07
n-Nonadecane	19	19.00	268.53		0.777	4.99E-07		**10.74**	**9.33E+08**	330	**3.10E-02**	**3.58E-06**			
n-Eicosane	20	20.00	282.6	3E-07	0.788	2.23E-07	8.00E+01	**11.27**	**2.54E+09**		**3.10E-02**	**3.61E-06**		3.25E-09	2.54E-09
n-Heneicosane	21	21.00													
n-Hexacosane	26	26.00	366.7	1E-10			8.10E+00								

Table 3. *Continued*

Compound	Carbons	EC	MW	S(mg/L)	SG	VP(atm)	H(c/c)	log K_{ow}	K_{oc}	BP(C)	D air(cm^2/s)	Dw(cm^2/s)	CAS #	LF	VF
STRAIGHT CHAIN ALKENES/ALKYNES															
trans-2-Butene	4	4.10						2.31	1.14E+02						
cis-2-Butene	4	4.25						2.33	1.18E+02						
1-Pentene	5	4.39	70.14	148	0.6405	8.39E-01	1.63E+01	2.80	2.88E+02	30	8.29E-02	8.72E-06	109-67-1	1.52E-02	6.45E-03
trans-2-Pentene	5	5.08						2.80	2.88E+02	36.9					
1-Pentyne	5	5.13	68.13	1570	0.6901	5.68E-01	1.01E+00	2.12	7.97E+01	40.18	8.37E-02	9.14E-06	627-19-0	8.08E-02	2.15E-03
cis-2-Pentene	5	5.16	70.134	203	0.6556	6.50E-01	9.20E+00	2.20	9.27E+01	36.9			627-20-3	3.43E-02	
1-Hexene	6	5.30	84.16	50	0.6732	2.45E-01	1.68E+01	3.39	8.77E+02	63.4	7.19E-02	7.91E-06	592-41-6	7.23E-03	2.76E-03
trans-3-Hexene	6	6.02						3.39	8.77E+02						
cis-3-Hexene	6	6.03						3.39	8.77E+02						
trans-2-Hexene	6	6.05						3.39	8.77E+02						
1-Hexyne	6	6.09	82.15	360	0.7155	1.79E-01	1.67E+00	2.73	2.52E+02	71.5			693-02-7	2.90E-02	
cis-2-Hexene	6	6.14						3.39	8.77E+02						
cis-3-Heptene	7	7.01						3.99	2.72E+03						
trans-2-Heptene	7	7.05	98.19	15	0.7012	6.37E-02	1.70E+01	3.99	2.72E+03	95.7			14686-13-6	2.76E-03	
1-Octene	8	7.89	112.1	2.7	0.7149	2.29E-02	3.89E+01	4.57	8.14E+03	121.3	8.37E-02	6.63E-06	111-66-0	9.45E-04	9.69E-04
1-Nonene	9	8.69	126.24	1.12	0.7292	7.03E-03	3.24E+01	5.15	2.43E+04	146.88	8.37E-02	5.76E-06	124-11-8	3.33E-04	2.84E-04
1-Decene	10	9.91	140.27	0.1	0.7408	2.13E-03	1.22E+02	5.31	3.29E+04	170.5	8.37E-02	5.36E-06	872-05-9	2.37E-04	7.66E-04
Propylene	3							1.77	4.12E+01						
Tridecene	13							7.03	8.46E+05						

Table 3. *Continued*

Compound	Carbons	EC	MW	S(mg/L)	SG	VP(atm)	H(c/c)	log K_{ow}	K_{oc}	BP(C)	D air(cm^2/s)	Dw(cm^2/s)	CAS #	LF	VF
BRANCHED CHAIN ALKANES															
Isobutane	4	3.67	58.13	48.9	0.5571	3.52E+00	4.86E+01	2.80	**2.88E+02**	-11.7	9.11E-02		78-78-4	7.95E-03	
Neopentane	5	4.32	72.15	33.2	0.591	1.70E+00	8.87E+01	3.11	**5.17E+02**	9.5	**8.10E-02**	**8.66E-06**	109-66-0	4.39E-03	9.94E-03
Isopentane	5	4.75	72.15	13.8	0.6193	9.04E-01	1.93E+02	**3.21**	**6.24E+02**	27.8	8.17E-02		78-78-4	2.30E-03	
2,2-Dimethylbutane	6	5.37	86.17	18.4	0.6492	4.20E-01	8.05E+01	3.82	**1.97E+03**	49.74	**7.20E-02**	**7.78E-06**	75-83-2	2.57E-03	4.69E-03
2,3-Dimethylbutane	6	5.68	86.17	19.1	0.6616	3.16E-01	5.83E+01	3.85	**2.09E+03**	58	**7.10E-02**	**7.80E-06**	79-29-8	2.76E-03	3.60E-03
2-Methylpentane	6	5.72	86.17	13.8	0.6532	2.78E-01	7.11E+01	**3.74**	**1.70E+03**	60.27	7.11E-02		107-83-5	2.96E-03	
3-Methylpentane	6	5.85	86.17	12.8	0.6643	2.50E-01	6.87E+01	3.60	**1.30E+03**	83.28	**7.00E-02**	**7.68E-06**	96-14-0	3.50E-03	5.30E-03
2,2-Dimethylpentane	7	6.25	100.21	4.4	0.6739	1.38E-01	1.29E+02	**4.14**	**3.61E+03**	79.2	**6.50E-02**	**7.10E-06**	590-35-2	1.48E-03	3.89E-03
2,4-Dimethylpentane	7	6.31	100.21	4.06	0.6727	1.29E-01	1.30E+02	**4.14**	**3.61E+03**	80.5	**6.40E-02**	**7.07E-06**	108-08-7	1.47E-03	3.87E-03
2,2,3-Trimethylbutane	7	6.36	100.21	4.38	0.6901	1.35E-01	1.26E+02	**4.03**	**2.93E+03**	80.9	**6.60E-02**	**7.30E-06**	464-06-2	1.70E-03	4.44E-03
3,3-Dimethylpentane	7	6.55	100.21	5.94	0.6936	1.08E-01	7.45E+01	**4.14**	**3.61E+03**	86.06	**6.50E-02**	**7.12E-06**	562-49-2	1.74E-03	2.65E-03
2-Methylhexane	7	6.68	100.21	2.54	0.6786	8.67E-02	1.40E+02	3.16	**5.68E+02**	90	1.87E-01	7.10E-06	591-76-4	3.04E-03	2.51E-02
2,3-Dimethylpentane	7	6.69	100.21	5.25	0.6951	9.06E-02	7.07E+01	**4.14**	**3.61E+03**	89.9	**6.60E-02**	**7.35E-06**	565-59-3	1.76E-03	2.58E-03
3-Methylhexane	7	6.76	100.21	3.3	0.6871	8.10E-02	1.01E+02	**4.27**	**4.62E+03**	92	**6.50E-02**	**7.23E-06**	589-34-4	1.34E-03	2.76E-03
3-Ethylpentane	7	6.86	100.21					**4.27**	**4.62E+03**		**6.40E-02**	**7.10E-06**			
2,2,4-Trimethylpentane	8	6.89	114.23	2.44	0.6919	6.47E-02	1.24E+02	**4.54**	**7.69E+03**	99.2	**6.00E-02**	**6.59E-06**	540-84-1	8.61E-04	2.02E-03
2,2-Dimethylhexane	8	7.25						**4.67**	**9.83E+03**		**5.90E-02**	**6.50E-06**			
2,2,3,3-Tetramethylbutane	8	7.30	114.23		0.8242	2.74E-02		**4.40**	**5.90E+03**	106.5	**6.00E-02**	**6.65E-06**	594-82-1		
2,5-Dimethylhexane	8	7.36	114.23		0.694			**4.67**	**9.83E+03**		**5.90E-02**	**6.47E-06**			
2,2,3-Trimethylpentane	8	7.37						**4.54**	**7.69E+03**		**6.20E-02**	**6.89E-06**			
2,4-Dimethylhexane	8	7.38	115.67		0.705			**4.67**	**9.83E+03**	109	6.17E-02				
3,3-Dimethylhexane	8	7.45						**4.67**	**9.83E+03**		**6.10E-02**	**6.82E-06**			

Table 3. *Continued*

Compound	Carbons	EC	MW	S(mg/L)	SG	VP(atm)	H(c/c)	log K_{ow}	K_{oc}	BP(C)	D air(cm^2/s)	Dw(cm^2/s)	CAS #	LF	VF
2,3,4-Trimethylpentane	8	7.55	114.23	2	0.7191	3.55E-02	8.30E+01	**4.54**	**7.69E+03**	113.4	**6.00E-02**	**6.65E-06**	565-75-3	9.21E-04	1.44E-03
2,3,3-Trimethylpentane	8	7.58						**4.54**	**7.69E+03**		**6.00E-02**	**6.71E-06**			
2,3-Dimethylhexane	8	7.65	115.67	0.13	0.72			**4.67**	**9.83E+03**	116	**5.90E-02**	**6.59E-06**			
2-Methyl-3-ethylpentane	8	7.66						**4.54**	**7.40E+03**		6.10E-02	**6.82E-06**			
2-Methylheptane	8	7.71	114.23	0.85	0.698	2.57E-02	1.41E+02	4.80	**1.26E+04**	117.6	**5.80E-02**	**6.42E-06**	562-27-6	5.61E-04	1.45E-03
4-Methylheptane	8	7.72						4.80	**1.26E+04**		**5.90E-02**	**6.52E-06**			
3,4-Dimethylhexane	8	7.74						**4.67**	**9.83E+03**		**5.90E-02**	**6.61E-06**			
3-Methylheptane	8	7.78	114.2	0.792	0.708	2.58E-02	1.52E+02	4.80	**1.26E+04**	115	**6.00E-02**	**6.63E-06**		5.55E-04	1.59E-03
3-Ethylhexane	8	7.79						**4.80**	**1.26E+04**		**6.00E-02**	**6.71E-06**			
2,2,5-Trimethylhexane	9	7.87	128.26	1.15	0.7072	2.18E-02	9.95E+01	**5.06**	**3.48E+04**	124	**5.50E-02**		3522-94-9	2.27E-04	
2,2,4-Trimethylhexane	9	7.93						**5.06**	**2.05E+04**		6.08E-02				
2,4,4-Trimethylhexane	9	8.07						**5.06**	**2.05E+04**						
2,3,5-Trimethylhexane	9	8.24						**5.06**	**2.05E+04**						
2,2-Dimethylheptane	9	8.28	128.26		0.71			5.20	**2.67E+04**	130					
2,4-Dimethylheptane	9	8.34						**5.20**	**2.67E+04**						
3,3-Dimethylheptane	9	8.42						**5.20**	**2.67E+04**						
2,5-Dimethylheptane	9	8.47	128.26		0.715			5.20	**2.67E+04**	136					
2,6-Dimethylheptane	9	8.47						**5.20**	**2.67E+04**						
3,4-Dimethylheptane	9	8.52						**5.20**	**2.67E+04**						
2,3-Dimethylheptane	9	8.54						**5.20**	**2.67E+04**						
4-Ethylheptane	9	8.59						**5.32**	**3.35E+04**						
4-Methyloctane	9	8.71	128.3	0.115	0.7199	8.90E-03	4.06E+02	5.32	**3.35E+04**	142.4				2.08E-04	
3-Ethylheptane	9	8.77						**5.32**	**3.35E+04**						

Table 3. *Continued*

Compound	Carbons	EC	MW	S(mg/L)	SG	VP(atm)	H(c/c)	log K_{ow}	K_{oc}	BP(C)	D air(cm^2/s)	Dw(cm^2/s)	CAS #	LF	VF
3-Methyloctane	9	8.78	128.3	1.42	0.714	8.12E-03		5.32	**3.35E+04**	143			2216-33-3		
2,6-Dimethyloctane	10	9.32						**5.85**	**9.12E+04**						
2-Methylnonane	10	9.72						**5.85**	**9.12E+04**						
3-Methylnonane	10	9.78						**5.85**	**9.12E+04**						
2,2,4-Trimethylheptane	10							**5.60**	**5.69E+04**						
3,3,5-Trimethylheptane	10							**5.60**	**5.69E+04**						
4-Methylnonane	10							**5.85**	**9.12E+04**						
2,4,6-Trimethylheptane	10							**5.60**	**5.69E+04**						
4-Methyldecane	11							**6.38**	**2.48E+05**						
2-Methyldecane	11							**6.38**	**2.48E+05**						
2-Methylundecane	12							**6.78**	**5.28E+05**						
2,6-Dimethyldecane	12							**6.78**	**5.28E+05**						
2,6-Dimethylundecane	13							**7.31**	**1.44E+06**						
2,4-Dimethyl-3-ethylpentane	9							**5.06**	**2.05E+04**						
2-Methyl-3-ethylhexane	9							**5.20**	**2.67E+04**						
2,3,3-Trimethylhexane	9							**5.06**	**2.05E+04**						
2-Methyloctane	9							**5.32**	**3.35E+04**						
BRANCHED CHAIN ALKENES															
3-Methyl-1-butene	5	4.57	70.14	130	0.6272	1.18E+00	2.21E+01	**2.66**	**2.21E+02**	20.1	**8.14E-02**		563-45-1	1.46E-02	
2-Methyl-1-butene	5	4.96						**2.66**	**2.21E+02**						
2-Methyl-2-butene	5	5.21						**2.66**	**2.21E+02**						
4-Methyl-cis-2-pentene	6	5.69						**3.19**	**6.01E+02**						
2,3-Dimethyl-1-butene	6	5.70						**3.30**	**7.40E+02**						

Table 3. *Continued*

Compound	Carbons	EC	MW	S(mg/L)	SG	VP(atm)	H(c/c)	log K_{ow}	K_{oc}	BP(C)	D air(cm^2/s)	Dw(cm^2/s)	CAS #	LF	VF
4-Methyl-trans-2-pentene	6	5.73						**3.13**	**6.01E+02**						
2-Methyl-1-pentene	6	5.89	84.16	78	0.6799	2.57E-01	1.13E+01	**3.20**	**6.12E+02**	60.7			76-20-3	1.04E-02	
2-Methyl-2-pentene	6	6.07						**3.13**	**6.01E+02**						
3-Methyl-cis-2-pentene	6	6.11						**3.13**	**6.01E+02**						
3-Methyl-trans-2-pentene	6	6.22						**3.13**	**6.01E+02**						
4,4-Dimethyl-trans-2 pentene	7	6.23						**3.59**	**1.28E+03**						
4,4-Dimethyl-cis-2-pentene	7	6.47						**3.59**	**1.28E+03**						
2,4-Dimethyl-1-pentene	7	6.48						**3.59**	**1.28E+03**						
3-Ethyl-2-pentene	7	7.07						**3.72**	**1.63E+03**						
2,3-Dimethyl-1-pentene	7							**3.59**	**1.28E+03**						
4,4-Dimethyl-1-pentene	7							**3.59**	**1.28E+03**						
CYCLOALKANES															
Cyclopentane	5	5.66	70.14	156	0.7454	4.18E-01	7.69E+00	3.00	**4.20E+02**	49.3	**8.57E-02**	**9.54E-06**	287-92-3	1.52E-02	3.15E-03
Methylcyclopentane	6	6.27	84.16	42	0.7486	1.81E-01	1.48E+01	3.37	**8.44E+02**	71.8	**7.52E-02**	**8.39E-06**	96-37-7	7.66E-03	2.69E-03
Cyclohexane	6	6.59	84.16	55	0.7786	1.25E-01	7.84E+00	3.44	**9.63E+02**	80.7	8.39E-02	9.10E-06	110-83-8	7.57E-03	1.57E-03
1,1-Dimethylcyclopentane	7	6.72						**3.83**	**2.01E+03**						
1-cis-3-Dimethylcyclopentane	7	6.82						**3.83**	**2.01E+03**						
1-trans-3-Dimethylcyclopentane	7	6.85						**3.83**	**2.01E+03**						
1-trans-2-Dimethylcyclopentane	7	6.87						**3.83**	**2.01E+03**						
n-Propylcyclopentane	8	7.10	112.21	2.04	0.7763	1.62E-02	3.64E+01	**4.37**	**5.58E+03**	101	6.18E-02	**7.00E-06**	2040-96-2	1.34E-03	9.53E-04
1-trans-2-trans-4-Trimethylcyclopentane	8	7.19	112.21		0.7665	5.30E-02		**4.35**	**5.37E+03**	103.5					
1-cis-2-Dimethylcyclopentane	7	7.21						**3.83**	**2.01E+03**						
Methylcyclohexane	7	7.22	98.19	14	0.7694	6.10E-02	1.75E+01	3.88	**2.21E+03**	100.9	6.66E-02	**7.66E-06**	108-87-2	3.32E-03	1.22E-03

Table 3. *Continued*

Compound	Carbons	EC	MW	S(mg/L)	SG	VP(atm)	H(c/c)	log K_{ow}	K_{oc}	BP(C)	D air(cm^2/s)	Dw(cm^2/s)	CAS #	LF	VF
1,1,3-Trimethylcyclopentane	8	7.25	112.21	3.73	0.7703	5.23E-02	6.44E+01	**4.35**	**5.37E+03**	104.9			4516-69-2	1.30E-03	
Ethylcyclopentane	7	7.34	98.19		0.7665	5.25E-02		**3.84**	**2.05E+03**	103.5	6.74E-02		1640-89-7		
1-trans-2-cis-3-Trimethylcyclopentane	8	7.51	112.21					**4.35**	**5.37E+03**						
1,1,2-Trimethylcyclopentane	8	7.67	112.21					**4.35**	**5.37E+03**						
1-cis-3-Dimethylcyclohexane	8	7.75	112.21					**4.39**	**5.79E+03**						
1-trans-4-Dimethylcyclohexane	8	7.80	112.2	3.84	0.763		3.56E+01	**4.39**	**5.79E+03**	119.4				1.30E-03	
1-trans-2-Dimethylcyclohexane	8	7.94	112.21	3.73	0.776	2.55E-02	3.14E+01	**4.39**	**5.79E+03**	124			6876-23-9	1.32E-03	
1-trans-3-Dimethylcyclohexane	8	7.99	112.21					**4.39**	**5.79E+03**			**6.88E-06**			
Ethylcyclohexane	8	8.38	112.2		0.7879			**4.40**	**5.90E+03**	131.8			1678-91-7		
1,1,3-Trimethylcyclohexane	9	8.45	126.24	1.77	0.7664	1.46E-02	4.26E+01	**4.91**	**1.55E+04**	138.94		**6.77E-06**	3073-66-3	5.12E-04	
Pentylcyclopentane	10	10.37	140.26	0.115	0.7912	1.50E-03	7.48E+01	**5.39**	**3.83E+04**	180			3741-00-2	2.09E-04	
1-Methyl-cis-2-ethylcyclopentane	8							**4.36**	**5.47E+03**						
1-Methyl-trans-3-ethylcyclopentane	8							**4.36**	**5.47E+03**						
1-trans-2-trans-4-Trimethylcyclohexane	8							**4.91**	**1.55E+04**						
1-Methyl-2-ethylcyclohexane	9							**4.92**	**1.58E+04**						
1-Methyl-3-ethylcyclohexane	9							**4.92**	**1.58E+04**						
1,3,5-Trimethylcyclohexane	9							**4.91**	**1.55E+04**						
1-Methyl-4-ethylcyclohexane	9							**4.92**	**1.58E+04**						
n-Propylcyclohexane	9							**4.93**	**1.61E+04**						
n-Butylcyclohexane	10							**5.46**	**4.37E+04**						
Hexylcyclohexane	12							**6.52**	**3.23E+05**						
Heptylcyclohexane	13							**7.12**	**1.00E+06**						

Table 3. *Continued*

Compound	Carbons	EC	MW	S(mg/L)	SG	VP(atm)	H(c/c)	log K_{ow}	K_{oc}	BP(C)	D air(cm^2/s)	Dw(cm^2/s)	CAS #	LF	VF
1-trans-2-cis-4-Trimethylcyclopentane	8		112.21					**4.35**	**5.37E+03**						
Isopropylcyclopentane	8		112.21					**4.37**	**5.58E+03**						
CYCLOALKENES															
Cyclopentene	5	5.55	68.12	535	0.772	5.00E-01	2.61E+00	**2.25**	**1.02E+02**	44			142-29-0	5.55E-02	
3-Methylcyclopentene	6	6.10						**2.77**	**2.72E+02**						
Cyclohexene	6	6.74	82.15	213	0.8102	1.17E-01	1.84E+00	2.86	**3.22E+02**	82.98	7.82E-02		110-83-8	2.31E-02	

approach for estimating partitioning. The chemical structure is separated into fragments and each fragment is assigned a weighting factor. The sum of the weighting factors is the log K_{ow}. ClogP is available in EPA programs (ASTER; Russom et al., 1991) and commercial risk assessment packages (Riskpro; GSC, 1990) and is widely used. Note that there are other methods published in the literature and as modules in commercially available software packages. The specific methods used are not always clearly stated.

Sangster (1989) performed a comprehensive review of log K_{ow} values for more than 50 petroleum hydrocarbons (other chemical classes were considered as well). Results of his analysis were a "best guess" log K_{ow} value and a numerical estimate (determined qualitatively) of error around the best guess value. Sangster's values were used in the Mackay et al. (1993) analysis of chemical properties. Errors range from 0.1 to 1.0 (on a log scale) with error typically increasing with chemical molecular size. This indicates that there is a greater uncertainty in K_{ow} estimates for the higher molecular weight compounds. The overall average error was 0.30 log units. A similar study (discussed in Mackay et al., 1993) reported the potential for errors in K_{ow} estimation and experiments for isobutene, cyclohexane, propane, indene, fluorene, anthracene, pyrene, neopentane, and styrene. This average error was 0.06 log units and the maximum error was 0.16 log units.

According to Lyman et al. (1990), estimates of log K_{ow} >6 are likely to be overestimated, possibly by one or more log units. Less conservatively, Mackay et al. (1993) state that calculated log K_{ow} values greater than 7 should be considered suspect and that calculated or experimental values greater than 8 should be treated with extreme caution. It is recognized that some of the high (>7) log K_{ow} values used in this study may be inaccurate, but they are useful in observing the trends in the transport characteristics of compounds with increasing molecular weight within an homologous series (hydrocarbon classes based on common molecular structures, e.g., cycloalkanes).

3.2.2.2 *Organic Carbon Partition Coefficient, K_{oc}*

Because only a limited amount of data was available in the literature, all K_{oc} values were estimated from K_{ow}, recorded as the average of the set of values calculated for a compound using the following equations obtained from the literature. Differences among the models would not result in order of magnitude variations of K_{oc} and thus were deemed adequate for this analysis, the goal of which is to determine fractions of TPH based on fate and transport considerations. More accurate K_{oc} values could be obtained using laboratory techniques. Table 4 presents the equations and chemical classification for which the estimation methods are most appropriate (Lyman et al., 1990; Olsen and Davis, 1990).

These equations were chosen from numerous studies in which correlations were developed between K_{ow} and K_{oc} for a variety of organic chemicals. The equations were chosen from the studies which best represented the compounds of concern here.

The uncertainty in the K_{oc} values estimated from the various equations may arise from a number of factors, including analytical method errors, uncertainty in the input data (as discussed above regarding K_{ow}), variability in environmental factors (e.g., pH, temperature, salinity), and errors resulting from extrapolation based on

Table 4. Estimation Equations for Log K_{oc}

Equation	Basis	Reference	Equation #
Log K_{oc} = 0.544 log K_{ow} + 1.377	Wide range of chemicals, mostly pesticides	Lyman et al., 1990	**(16)**
Log K_{oc} = 0.937 log K_{ow} - 0.006	Aromatics, polynuclear aromatics, herbicides	Lyman et al., 1990	**(17)**
Log K_{oc} = 1.00 log K_{ow} - 0.21	Mostly aromatics and polynuclear aromatics	Lyman et al., 1990	**(18)**
Log K_{oc} = 0.63 log K_{ow}		Olsen and Davis, 1990	**(19)**
Log K_{oc} = 0.989 log K_{ow}- 0.346		Olsen and Davis, 1990	**(20)**

assumptions of a linear isotherm and reversible adsorption. Method errors are typically less than one order of magnitude, but in some rare occurrences they can amount to two orders of magnitude. Log K_{oc} values may be as high as 7 log units (Lyman et al., 1992). It is recognized, as with K_{ow}, that some log K_{oc} values greater than 7 used in this study may be inaccurate, but they are nevertheless useful in observing the trends in transport with increasing molecular weight.

3.2.2.3 Diffusivity in Air, D_{air}

D_{air}, the diffusion coefficient in air, is a function of the Boltzmann's constant, chemical molecular weight, collision integral, and the characteristic length of the molecules. Its units are cm^2/s. The estimation method used for D_{air} is from McCabe et al. (1985) and is based on the following, where "a" is air and "b" is the chemical of interest.

$$D_{air} = \frac{0.01498T^{1.81}\left(\frac{1}{M_a} + \frac{1}{M_b}\right)^{0.5}}{P(T_{ca}T_{cb})^{0.01405}(V_{ca}^{0.4} + V_{cb}^{0.4})^2} \qquad (21)$$

where:

T = temperature [K]
M = molecular weight [g/mole]
P = pressure [atm]
T_c = critical temperature [K]
V_c = critical molar volume [cm^3/mole]

Molar volume data were obtained from Reid et al. (1977). When these data were unavailable, critical molar volume was estimated from methods presented in Smith and Van Ness (1975).

The accuracy of the above estimation method was not stated. However, all diffusivity values, both measured and estimated were within an order of magnitude of each other. Errors in diffusivity values should have little effect on volatilization factors as calculated in this study.

3.2.3 *Summary of Literature Review*

To summarize, more than 260 specific chemicals were evaluated. Their boiling points range from -5°C to 350°C, and carbon number ranges from 5 to greater than 20. Both aromatic and aliphatic compounds are well represented in the form of 11 homologous series. These are:

- straight chain alkanes
- straight chain alkenes
- straight chain alkynes
- branched chain alkanes
- branched chain alkenes
- cycloalkanes
- cycloalkenes
- alkyl benzenes (including benzene)
- naphtheno benzenes
- alkyl naphthalenes (including naphthalene)
- polynuclear aromatics

4.0 DEVELOPMENT OF FRACTION-SPECIFIC PHYSICAL-CHEMICAL PROPERTIES USING PARTITIONING AND TRANSPORT EQUATIONS

4.1 USE OF RBCA SCREENING-LEVEL MODELS

Equations for Leaching Factor (LF) and Volatilization Factor (VF) are presented in Section 2. The values of these factors were used to group chemicals into specific fractions. As discussed previously, LF and VF are dependent on both chemical-specific and site-specific properties. The values of LF and VF are based on equations and assumptions specific to this project and should not be referenced or used for other modeling purposes.

4.1.1 Chemical-Specific Properties

Data were not available for the physical-chemical properties of all 260 compounds listed in Table 3. About 180 chemicals had available data for all properties, and most had data for several properties. Chemical properties have been discussed in detail in previous sections.

Table 5. Site Parameters

Symbol	Parameter	Units	Value
ρ_s	Soil bulk density	g/cm^3	1.7
θ_{ws}	Volumetric water content in vadose zone soils	cm^3/cm^3	0.12
θ_{as}	Volumetric air content in vadose zone soils	cm^3/cm^3	0.26
U_{gw}	Groundwater Darcy velocity	cm/y	2500
δ_{gw}	Groundwater mixing zone thickness	cm	200
I	Infiltration rate of water through soil	cm/y	30
W	Width of source area parallel to wind direction or groundwater flow	cm	1500
f_{oc}	Fraction of organic carbon in soil	g/g	.01
U_{air}	Wind speed in mixing zone	cm/s	225
δ_{air}	Ambient air mixing zone height	cm	200
L_s	Depth to subsurface soil sources	cm	100
θ_t	Total soil porosity	cm^3/cm^3	.38

4.1.2 Site Properties

This analysis is designed to develop fraction-specific physical-chemical properties for typical petroleum hydrocarbon release sites. Thus, the default site values in the RBCA standard were used. These values are presented in Table 5. Variation in site parameters is not expected to change the general conclusions of the study (definition of fractions) nor the representation of the physical-chemical properties of each fraction. Use of different site values would change all the calculated LF and VF values in a consistent direction resulting in the same definition of fractions. Values for LF and VF presented in Table 3 are based on these site parameters.

4.1.3 Comparison of Chemical-Specific Leaching and Volatilization Factors

The leaching factor and volatilization factor for each of the 180 chemicals are presented in Figures 3 and 4 as a function of equivalent carbon number (EC). Each homologous series was identified separately and trends identified. It is very apparent that aliphatics and aromatics behave differently in the environment, as expressed in the leaching factor. Aromatic chemicals are typically more soluble in water and slightly less volatile in air than aliphatic chemicals of similar equivalent carbon number. Additionally, the range of values spans many order of magnitude (9 - 12), thus providing the basis that order of magnitude differentiation of fractions is sufficient.

Table 6. Fractions Defined in this Study

Range of Equivalent Carbon Number, EC	Classification
5 - 6	Aliphatic
>6 - 8	Aliphatic
>8 - 10	Aliphatic
>10 - 12	Aliphatic
>12 - 16	Aliphatic
>16 - 35	Aliphatic
Benzene (6.5)	Aromatic
Toluene (7.6)	Aromatic
>8 - 10	Aromatic
>10 - 12	Aromatic
> 12 - 16	Aromatic
>16 - 21	Aromatic
>21 - 35	Aromatic

4.2 SELECTION OF FRACTIONS

Modeling the transport of chemicals in the subsurface as individual compounds is intractable for petroleum hydrocarbon distillates and crude oils because they contain large numbers of individual compounds. It is therefore desirable to group hydrocarbon compounds into a small number of fractions having similar transport properties to simplify modeling. Figures 3 and 4 were used to group compounds having similar LF and VF values ranging one order of magnitude for both aromatics and aliphatics. This is a reasonable level of accuracy, given the simplifying assumptions and uncertainty inherent in modeling the behavior of hydrocarbons in soils and is consistent with other approaches dealing with complex mixtures (Bischoff, et al., 1991; Peterson, 1994). The fractions are represented by the range in equivalent carbon number, EC. The fractions defined by this study are shown in Table 6.

Note that benzene and toluene have been identified as separate fractions. This is primarily for convenience, because benzene is likely to be evaluated as a carcinogen in addition to the noncancer evaluation described in this report. Ethylbenzene and the xylenes are in the >8 - 10 aromatic fraction. For ease of application, it may be useful to compress or combine fractions or evaluate only those that represent a potential for health risk at a site. If this is done, however, it is important to retain the appropriate fate and transport characteristics as they will drive an exposure assessment.

When specifying this protocol to an analytical lab, the resultant data should be presented in terms of "equivalent carbon number" (not carbon number). Leaching factors and volatilization factors presented for those petroleum hydrocarbons that fall in each fraction are displayed in Figures 5 and 6.

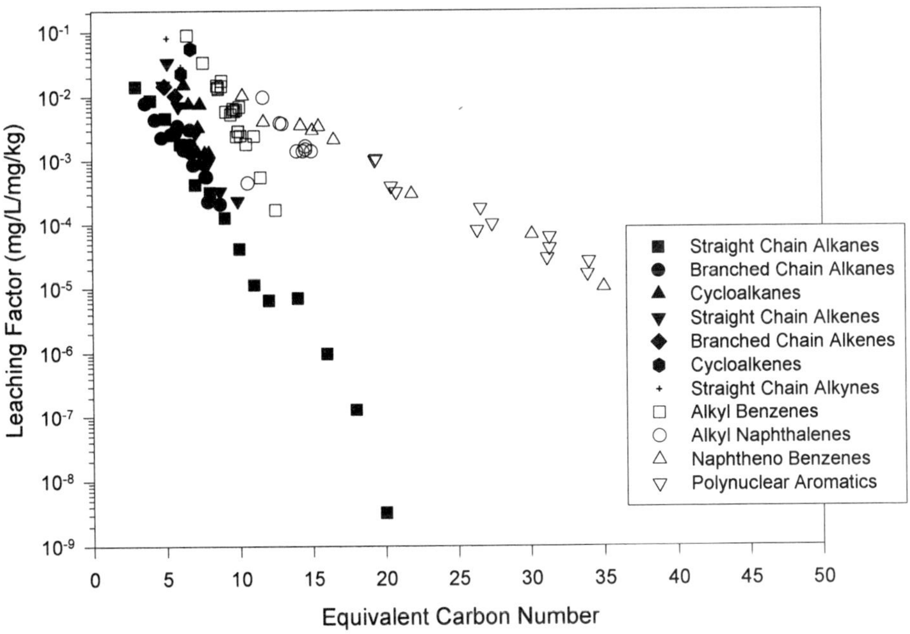

Figure 3. Leaching Factor by Homologous Series

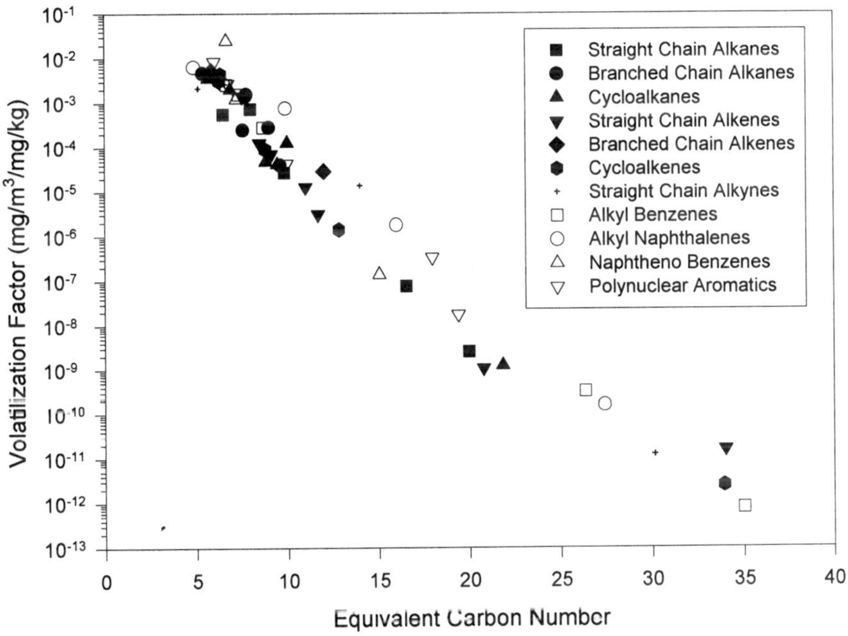

Figure 4. Volatilization Factor by Homologous Series

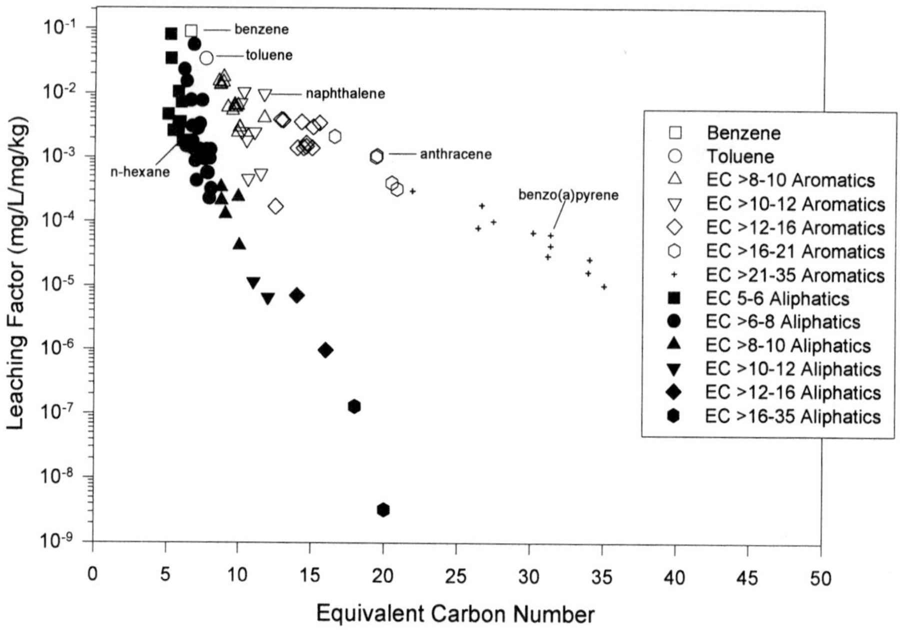

Figure 5. Leaching Factor By Representative Fractions

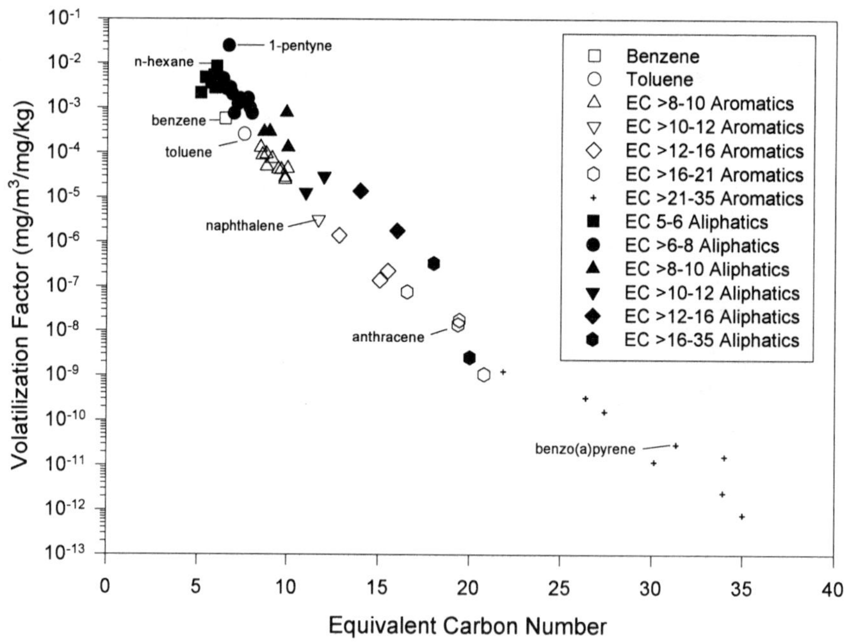

Figure 6. Volatilization Factor by Representative Fractions

4.3 ESTIMATION OF FRACTION PROPERTIES

There are several alternatives for estimating representative physical-chemical properties for each fraction. In this study, we evaluated several approaches using almost all available data. Physical-chemical properties of the alkynes were not included in the analyses because of the relatively small quantities found in petroleum mixtures and because these extreme values would significantly skew the data set.

A summary of the alternate approaches considered in the project is presented below:

1. **Simple Averaging:** In this case, all available property data within a fraction were averaged. The maximum and minimum values of each fraction were also investigated; however, it was decided that the average more closely represents the estimates based on the composition of gasoline.
2. **Composition-Based Averaging:** A weighted average of the available data for each fraction was computed based on the relative mass of each component in gasoline (other mixtures could have been used). It is acknowledged that molar average may have been more appropriate; however, because the range of molecular weights of the components in each fraction is relatively narrow compared to the molecular weight of gasoline, the difference is insignificant.
3. **Correlation to Relative Boiling Point Index:** Empirical relationships of the properties were developed based on a unifying characteristic (equivalent carbon number, EC) using all available data from Table 3. For each fraction, the EC value in the midpoint of the group served as the independent variable in the equation. This approach is similar to the work presented in Eastcott et al. (1988) and API (1992).

Each method yielded similar results (Table 7). It was determined that the correlations approach (No. 3) would be most useful, because if the definition of the fractions changed, new properties could be easily computed without having to resort and re-average the compound lists. Thus, in the remainder of this section we will discuss the results of the correlation method.

One important modification deals with the parameters developed for the Aliphatic EC>16-35 fraction. Due to the paucity of data on aliphatic chemicals with equivalent carbon numbers greater than 21 and the difficulty in accurately measuring values for these large molecules, the properties of the EC>16-35 fraction were based on extrapolation of those properties of compounds with equivalent carbon numbers up to 21. As both the lighter and heavier ends of the range within the fraction behave similarly in the environment (e.g., strongly binds to soil), the estimates should be adequate for performing exposure assessments.

Table 7. Comparison of Fraction-Specific Properties: Averaging, Weighted Averaging, and Correlations

Fraction	BP (°C)	EC	MW (g/mole)	S (mg/L)	VP (atm)	H (cm^3/cm^3)	log K_{oc}
			Properties Based on Averaging of Fractions				
ALIPHATICS							
EC 5-6	5.4E+01	5.6E+00	8.0E+01	1.0E+02	3.8E-01	4.1E+01	2.8E+00
EC >6-8	1.0E+02	7.0E+00	1.1E+02	1.6E+01	7.0E-02	7.7E+01	3.5E+00
EC >8-10	1.5E+02	8.8E+00	1.3E+02	6.9E-01	6.9E-03	1.6E+02	4.5E+00
EC >10-12	2.0E+02	1.1E+01	1.6E+02	5.3E-02	7.2E-04	1.6E+02	5.5E+00
EC >12-16	2.7E+02	1.5E+01	2.1E+02	3.5E-04	3.9E-05	1.6E+02	6.7E+00
EC >16-35	3.2E+02	2.0E+01	2.8E+02	1.5E-06	1.1E-06	1.1E+02	8.6E+00
AROMATICS							
EC 5-7	8.0E+01	6.5E+00	7.8E+01	1.8E+03	1.3E-01	2.2E-01	1.9E+00
EC >7-8	1.1E+02	7.6E+00	9.2E+01	5.2E+02	3.8E-02	2.7E-01	2.4E+00
EC >8-10	1.6E+02	9.3E+00	1.2E+02	1.1E+02	6.0E-03	4.2E-01	3.1E+00
EC >10-12	1.9E+02	1.1E+01	1.4E+02	3.0E+01	9.4E-04	3.4E-01	3.5E+00
EC >12-16	2.6E+02	1.4E+01	1.5E+02	9.3E+00	6.0E-05	9.7E-02	3.8E+00
EC >16-21	3.4E+02	1.9E+01	1.8E+02	5.6E-01	2.3E-06	9.9E-03	4.2E+00
EC >21-35	4.7E+02	2.9E+01	2.5E+02	2.9E-02	1.6E-08	8.2E-05	5.1E+00

Table 7. *Continued*

Properties Based on Weighted Averaging of Fractions (by Weight Percent in Gasoline)

Fraction	BP (°C)	EC	MW (g/mole)	S (mg/L)	VP (atm)	H (cm^3/cm^3)	log K_{oc}
ALIPHATICS							
EC 5-6	4.8E+01	5.3E+00	7.9E+01	3.6E+01	5.5E-01	9.7E+01	3.0E+00
EC >6-8	9.4E+01	6.9E+00	1.0E+02	1.2E+01	9.0E-02	8.6E+01	3.6E+00
EC >8-10	1.5E−02	9.1E+00	1.3E+02	1.2E-01	1.0E-02	2.9E+02	4.8E+00
EC >10-12	2.0E−02	1.1E+01	1.6E+02	3.0E-02	4.0E-04	1.5E+02	5.9E+00
EC >12-16	2.7E−02	1.5E+01	2.1E+02	3.5E-04	3.9E-05	1.6E+02	6.7E+00
EC >16-35	3.2E+02	2.0E+01	2.8E+02	1.5E-06	1.1E-06	1.1E+02	8.6E+00
AROMATICS							
EC 5-7	8.0E+01	6.5E+00	7.8E+01	1.8E+03	1.3E-01	2.3E-01	1.9E+00
EC >7-8	1.1E+02	7.6E+00	9.2E+01	5.2E+02	3.8E-02	2.7E-01	2.4E+00
EC >8-10	1.5E+02	9.3E+00	1.1E+02	1.3E+02	7.8E-03	2.7E-01	2.9E+00
EC >10-12	1.9E+02	1.1E+01	1.3E+02	4.7E+01	1.3E-03	3.4E-01	3.2E+00
EC >12-16	2.6E+02	1.4E+01	1.5E+02	9.3E+00	6.0E-05	9.7E-02	3.8E+00
EC >16-21	3.4E+02	1.9E+01	1.8E+02	5.6E-01	2.3E-06	9.9E-03	4.2E+00
EC >21-35	4.7E+02	2.9E+01	2.5E+02	2.9E-02	1.6E-08	8.2E-05	5.1E+00

Table 7. *Continued*

Properties Based on Correlations

Fraction	BP (°C)	EC	MW (g/mole)	S (mg/L)	VP (atm)	H (cm^3/cm^3)	log K_{oc}
ALIPHATICS							
EC 5-6	5.1E+01	5.5E+00	8.1E+01	3.6E+01	3.5E-01	4.7E+01	2.9E+00
EC >6-8	9.6E+01	7.0E+00	1.0E+02	5.4E+00	6.3E-02	5.0E+01	3.6E+00
EC >8-10	1.5E+02	9.0E+00	1.3E+02	4.3E-01	6.3E-03	5.5E+01	4.5E+00
EC >10-12	2.0E+02	1.1E+01	1.6E+02	3.4E-02	6.3E-04	6.0E+01	5.4E+00
EC >12-16	2.6E+02	1.4E+01	2.0E+02	7.6E-04	4.8E-05	6.9E+01	6.7E+00
EC >16-21	3.2E+02	1.9E+01	2.7E+02	2.5E-06	1.1E-06	8.5E+01	8.8E+00
AROMATICS							
EC 5-7	8.0E+01	6.5E+00	7.8E+01	2.2E+02	1.1E-01	1.5E+00	3.0E+00
EC >7-8	1.1E+02	7.6E+00	9.2E+01	1.3E+02	3.5E-02	8.6E-01	3.1E+00
EC >8-10	1.5E+02	9.0E+00	1.2E+02	6.5E+01	6.3E-03	3.9E-01	3.2E+00
EC >10-12	2.0E+02	1.1E+01	1.3E+02	2.5E+01	6.3E-04	1.3E-01	3.4E+00
EC >12-16	2.6E+02	1.4E+01	1.5E+02	5.8E+00	4.8E-05	2.8E-02	3.7E+00
EC >16-21	3.2E+02	1.9E+01	1.9E+02	6.5E-01	1.1E-06	2.5E-03	4.2E+00
EC >21-35	3.4E+02	2.8E+01	2.4E+02	6.6E-03	4.4E-10	1.7E-05	5.1E+00

4.3.1 Water Solubility [mg/L]

Figure 7 shows the correlation for water solubility as a function of equivalent carbon number. The greater solubility of aromatic hydrocarbons as compared to aliphatic hydrocarbons of similar equivalent carbon number is very evident, especially at high EC values. Also note that the variability of solubility around any given EC is bounded by about an order of magnitude. The resultant correlations are:

$$\log_{10} S = -0.55 \cdot EC + 4.5, \text{ for aliphatics } (R^2 = .94) \quad (22)$$

$$\log_{10} S = -0.21 \cdot EC + 3.7, \text{ for aromatics } (R^2 = .89) \quad (23)$$

The American Petroleum Institute (API, 1992) has developed correlations of solubility vs. boiling point for different chemical classes (Figure 8). These curves represent the changing nature of the compound as more aliphatic groups are added to an aromatic ring. Thus, compounds of mixed classification (alkyl benzenes) would demonstrate water solubilities between the two lines in Figure 7. The more alkylated the compound, the closer the water solubility would be to the aliphatic line.

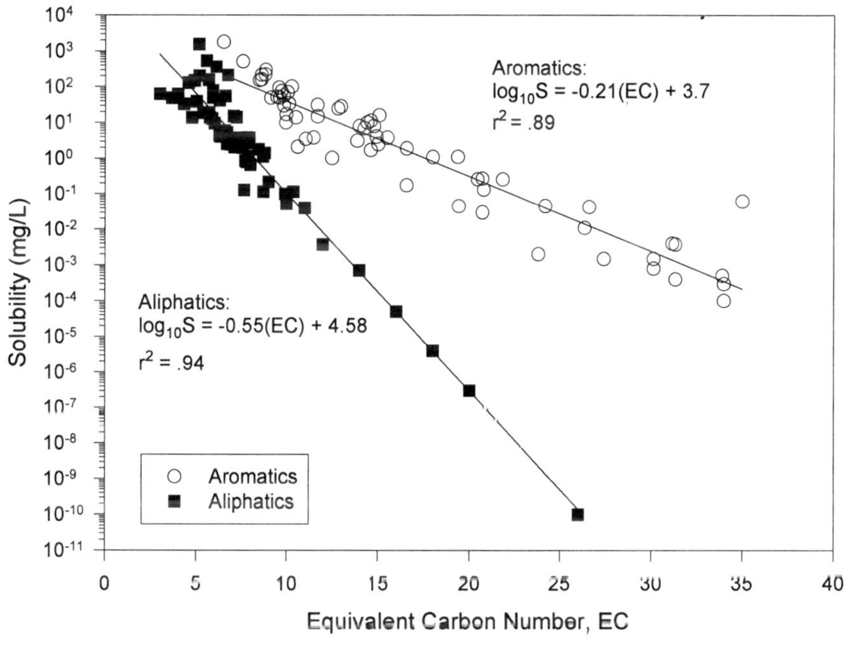

Figure 7. Solubility vs. Equivalent Carbon Number

4.3.2 Vapor Pressure [atm]

Figure 9 shows a plot of vapor pressure as a function of EC. There is very little difference between aliphatics and aromatics. This is due to the dependence of both vapor pressure and equivalent carbon number index on boiling point. Thus, correlations were not differentiated between aromatic and aliphatic classes, but separate curves were fit through data where EC is less than or equal to 12 and greater than 12. Vapor pressure values for equivalent carbon numbers greater than 25 were excluded due to the low confidence in the measured estimates of VP values less than 10^{-9} atm. This correlation is very similar to the API correlation (API, 1992) shown in Figure 10.

$$\log_{10} VP=-0.50 \cdot EC+2.3, \text{ for } EC>12 \ (R^2 = .99) \qquad (24)$$

$$\log_{10} VP=-0.36 \cdot EC+.72, \text{ for } EC>12 \ (R^2 = .96) \qquad (25)$$

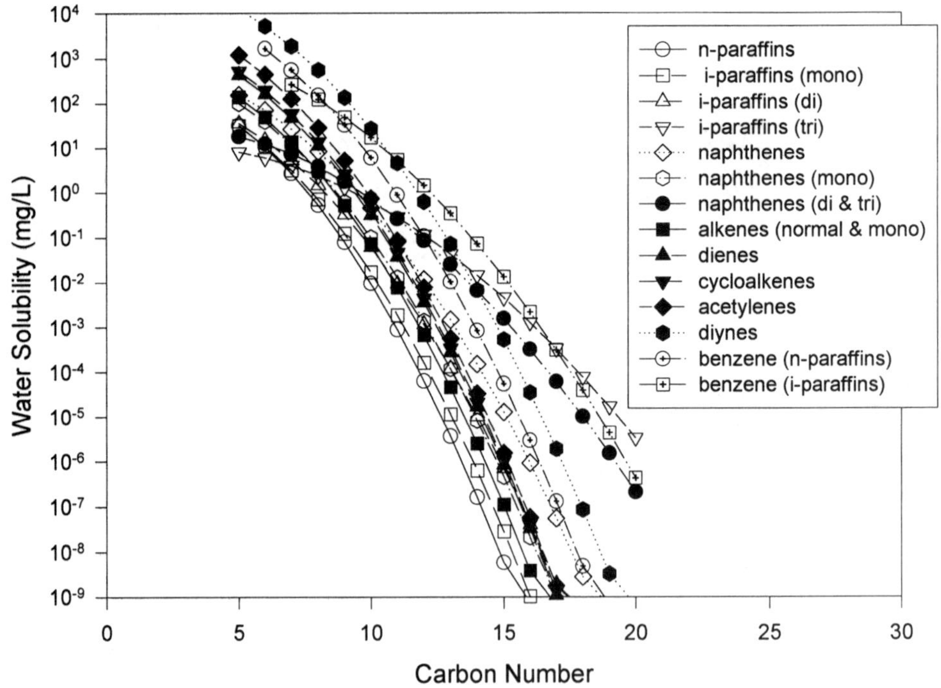

Figure 8. API Correlation for Water Solubility

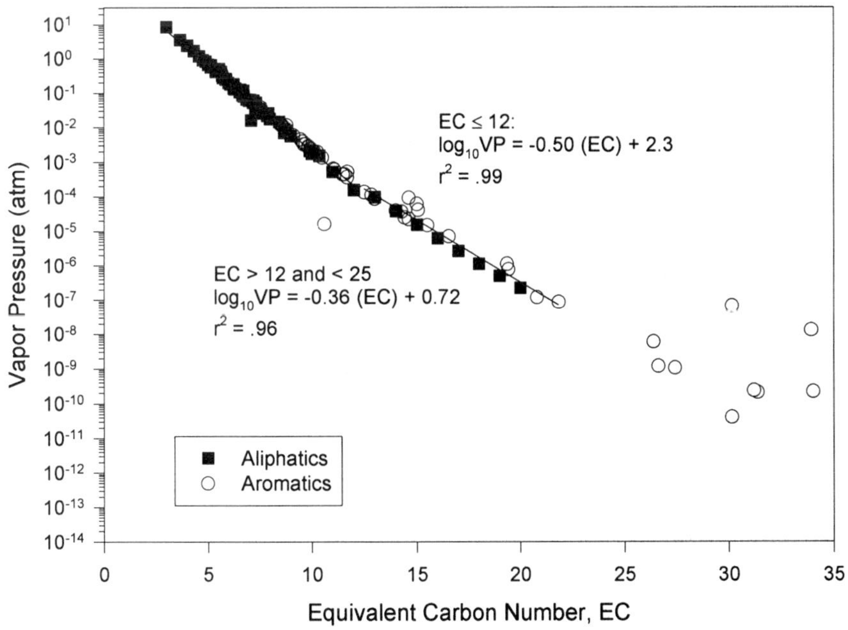

Figure 9. Vapor Pressure vs. Equivalent Carbon Number

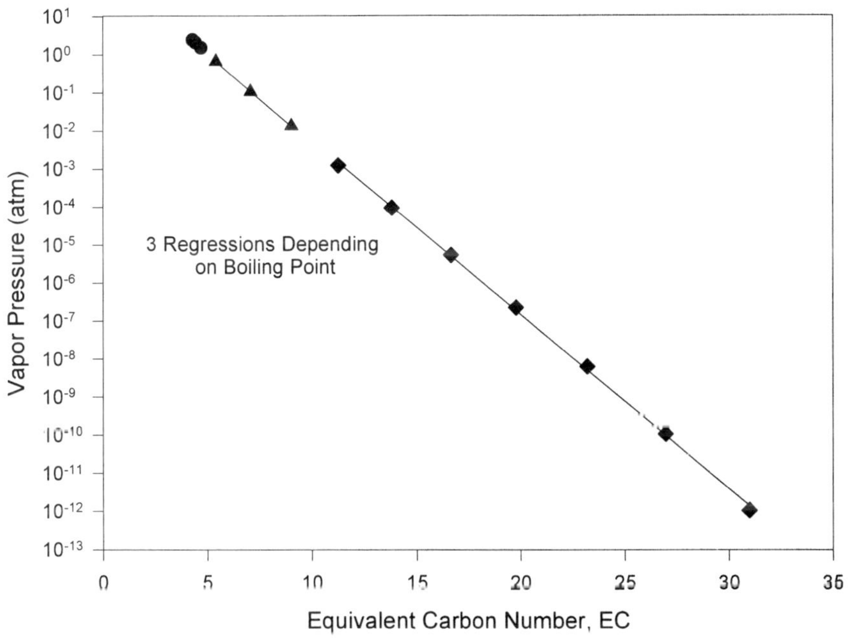

Figure 10. API Correlation for Vapor Pressure (Procedure 5A1.13)

4.3.3 Henry's Law Constant [cm³/cm³]

Figure 11 shows a plot of concentration-based Henry's law constant as a function of equivalent carbon number. There are several significant observations from this figure. First, there is a qualitative difference in the behavior of the aliphatic and aromatic hydrocarbons. The Henry's law constant decreases with increasing EC of the aromatics, while it increases for the aliphatics (although there is little confidence in the correlation fit to these data). More importantly, the Henry's law constants for the aliphatics are generally greater than 1.0, while those for the aromatics are generally less than 1.0. Thus, the aromatics prefer to be in the aqueous phase as compared to the aliphatics, especially at the higher EC values.

Fraction-specific values could be estimated from equations 26 and 27, or by using estimated molecular weights, solubilities, and vapor pressures and calculating a Henry's law constant. This latter approach, presented in Equation 28, is used in this study since it allows for internal consistency with other estimated values (i.e., S and VP).

$$\log_{10}H = 0.02 \cdot EC + 1.6, \text{ for aliphatics } (R^2 = .03) \quad (26)$$

$$\log_{10}H = -0.23 \cdot EC + 1.7, \text{ for aromatics } (R^2 = .84) \quad (27)$$

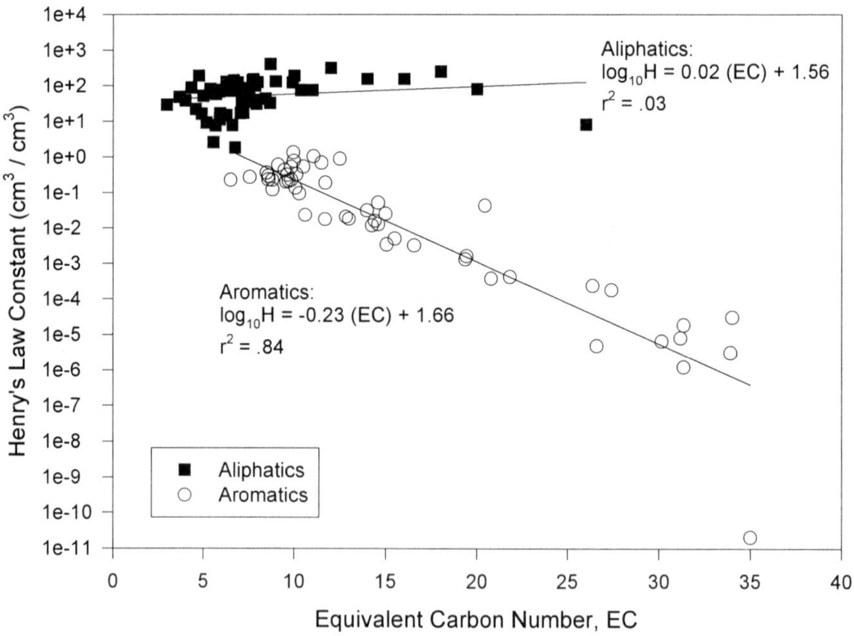

Figure 11. Henry's Law Constant vs. Equivalent Carbon Number

$$H = \frac{VP \cdot MW}{S \cdot R \cdot T} \quad (28)$$

All parameters have been previously defined.

4.3.4 Organic Carbon Partition Coefficient [mL/g]

A plot of log K_{oc} as a function of equivalent carbon number is shown in Figure 12. In this case, the aliphatics are more strongly absorbed onto the organic material in the soil as evidenced by the increase in log K_{oc} with increasing EC relative to that of the aromatics. Correlations were fit to the data as shown in the following equations.

$$\log_{10} K_{oc} = 0.45 \cdot EC + 0.43, \text{ for aliphatics } (R^2 = .94) \quad (29)$$

$$\log_{10} K_{oc} = 0.10 \cdot EC + 2.3, \text{ for aromatics } (R^2 = .81) \quad (30)$$

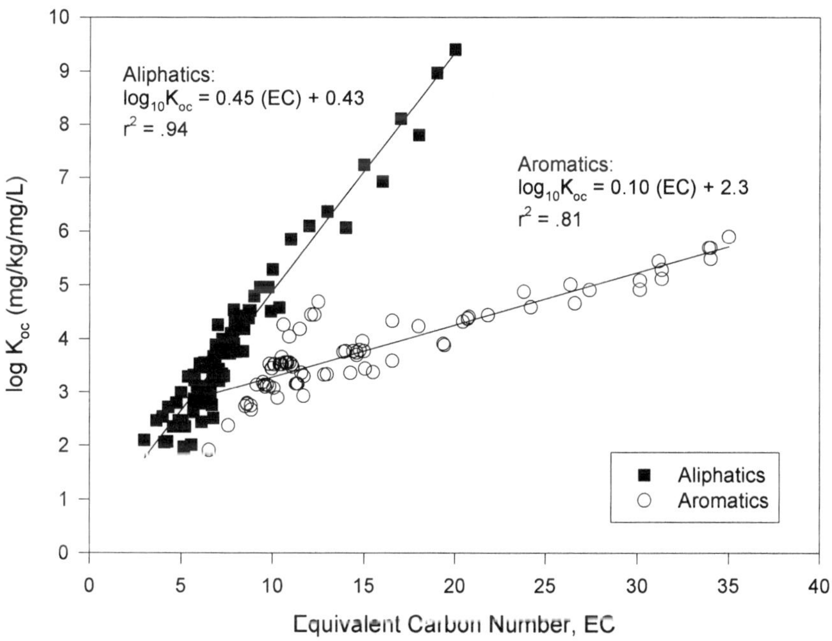

Figure 12. Sorption Coefficient vs. Equivalent Carbon Number

4.3.5 Diffusivity in Air and Water [cm²/s]

Figure 13 shows plots for the diffusivity in air and water as functions of equivalent carbon number. The important observation is that diffusivity (air or water) does not vary significantly from compound to compound. Thus, a conservative, reasonable assumption would be to set $D_{air} = 10^{-1}$ cm^2/sec and $D_{water} = 10^{-5}$ cm^2/sec for all fractions with little loss in accuracy.

4.4 PARTITIONING BEHAVIOR

Partitioning estimates, expressed in terms of a mass balance across media, are presented in Figures 14, 15, and 16. The phase weight fractions were calculated using the equations discussed in Section 2.1 for each chemical in Table 3 (where data were available).

The phase weight fraction presented is an estimate of the fraction of each hydrocarbon which will partition into each phase under equilibrium conditions (if no NAPL is present). Note that a significant fraction of the aliphatics is in the vapor phase while very little is in the aqueous phase at low EC values. Conversely, for aromatics, the significant mass is in the aqueous phase. At higher EC values (>10), most of the mass for either aliphatic or aromatic hydrocarbons is sorbed to the organic material in the soil.

4.5 SUMMARY OF FRACTION-SPECIFIC FATE AND TRANSPORT PROPERTIES

Using the empirical correlations presented above, fraction-specific properties were estimated. As stated earlier, the equivalent carbon number (EC) at the midpoint of each fraction was used as the independent variable in the equation. For example, the EC>10-12 aromatic fraction is represented by the equivalent carbon number of 11. Eleven is then entered in the above equations to calculate the EC>10-12 aromatic properties. Note that actual properties were used for the benzene and toluene fractions and that the properties for the EC >16-35 aliphatic fraction are based on only those chemicals in the 16-21 equivalent carbon range. The fraction-specific properties are summarized in Table 8. They could be used to estimate exposure-potential of the specific hydrocarbon fractions. For this project, we have selected the ASTM RBCA models to estimate exposure in a simple screening-level analysis. For cases where more complex fate and transport models are required, the properties provided in Table 8 can be used as input parameters and each fraction modeled individually. Additional fraction-specific fate and transport factors, such as biodegradation have not been determined as part of the TPHCWG work but should be considered in Tier II and III exposure assessments.

5.0 Validation of Fate and Transport Properties

Validation of fate and transport properties can be determined by both laboratory and field studies. Laboratory studies involve estimating specific chemical properties (e.g., water solubility) for hydrocarbon mixtures in bi-phasic systems and presenting the results in terms of the specific fractions. Additionally, leaching factors

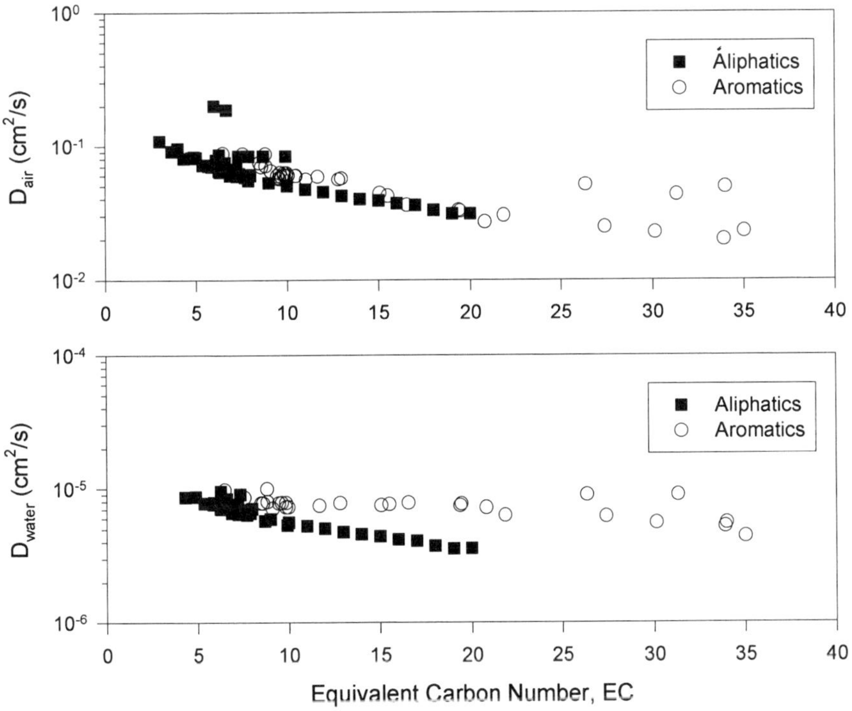

Figure 13. Diffusivity vs. Equivalent Carbon Number

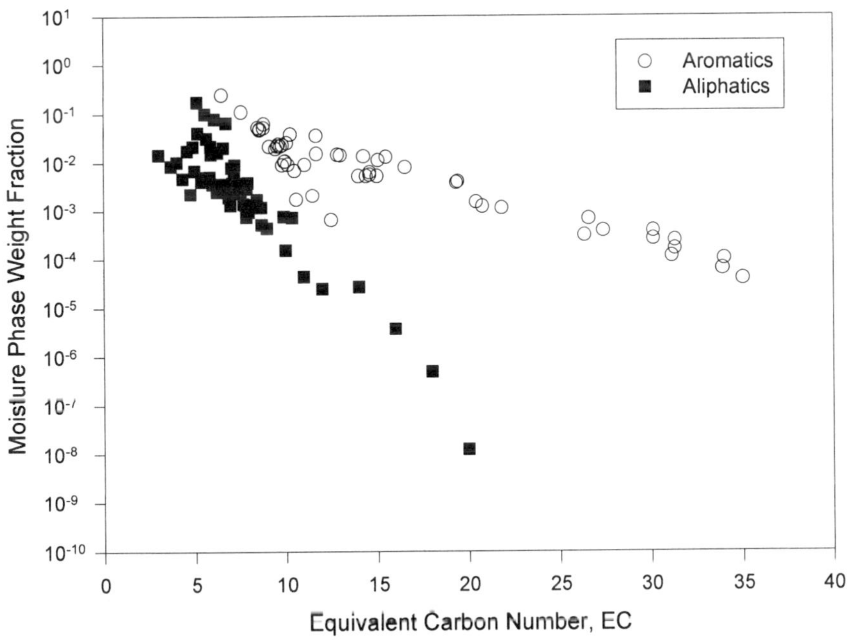

Figure 14. Partitioning to Moisture Phase

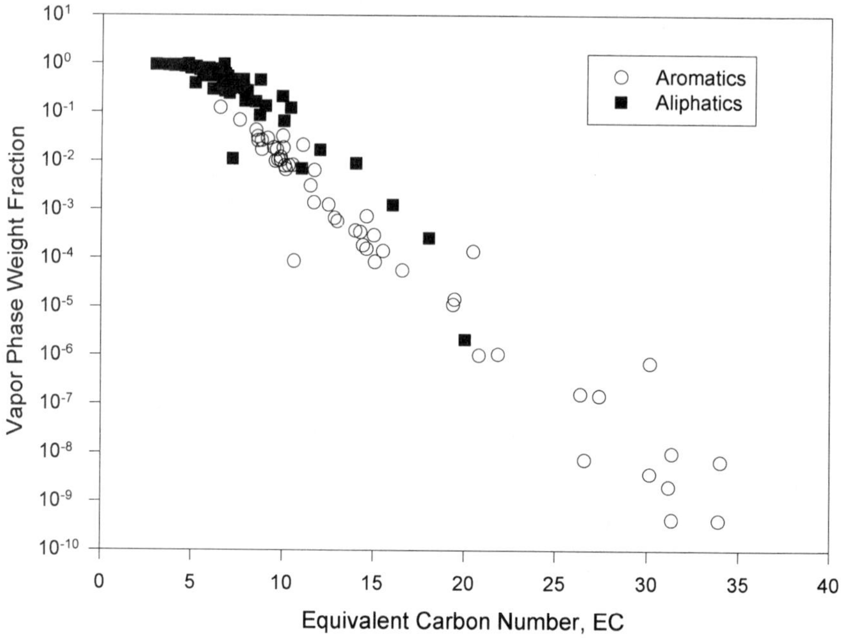

Figure 15. Partitioning to Vapor Phase

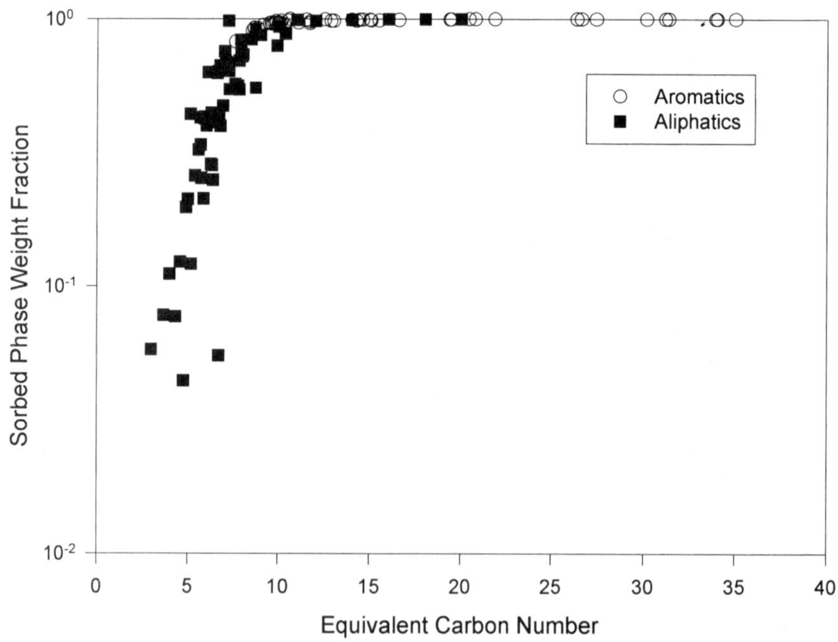

Figure 16. Partitioning to Sorbed Phase

Table 8. Fraction-Specific Physical-Chemical Properties

Fraction	BP (°C)	EC	MW (g/mole)	S (mg/L)	VP (atm)	H (cm^3/cm^3)[b]	log K_{oc}
ALIPHATICS							
EC 5-6	5.1E+01	5.5E+00	8.1E+01	3.6E+01	3.5E-01	3.3E+01	2.9E+00
EC >6-8	9.6E+01	7.0E+00	1.0E+02	5.4E+00	6.3E-02	5.0E+01	3.6E+00
EC >8-10	1.5E+02	9.0E+00	1.3E+02	4.3E-01	6.3E-03	8.0E+01	4.5E+00
EC >10-12	2.0E+02	1.1E+01	1.6E+02	3.4E-02	6.3E-04	1.2E+02	5.4E+00
EC >12-16	2.6E+02	1.4E+01	2.0E+02	7.6E-04	4.8E-05	5.2E+02	6.7E+00
EC >16-21	3.2E+02	1.9E+01	2.7E+02	2.5E-06	1.1E-06	4.9E+03	8.8E+00
AROMATICS							
Benzene (EC 5-7)[a]	8.0E+01	6.5E+00	7.8E+01	1.8E+03	1.3E-01	2.3E-01	1.9E+00
Toluene (EC >7-8)[a]	1.1E+02	7.6E+00	9.2E+01	5.2E+02	3.8E-02	2.7E-01	2.4E+00
EC >8-10	1.5E+02	9.0E+00	1.2E+02	6.5E+01	6.3E-03	4.8E-01	3.2E+00
EC >10-12	2.0E+02	1.1E+01	1.3E+02	2.5E+01	6.3E-04	1.4E-01	3.4E+00
EC >12-16	2.6E+02	1.4E+01	1.5E+02	5.8E+00	4.8E-05	5.3E-02	3.7E+00
EC >16-21	3.2E+02	1.9E+01	1.9E+02	6.5E-01	1.1E-06	1.3E-02	4.2E+00
EC >21-35	3.4E+02	2.8E+01	2.4E+02	6.6E-03	4.4E-10	6.7E-04	5.1E+00

[a] Actual values of benzene and toluene, not based on correlation.

[b] Calculated Henry's law constant based on vapor pressure, solubility, and molecular weight relationship.

and volatilization factors can be measured in the lab using tri-phasic systems (e.g., unsaturated soil columns). Field validation and/or application of the TPHCWG approach involves comparing measured concentrations in soil, groundwater, and soil gas with those predicted from models using fraction-specific fate and transport properties as input. This section will discuss, briefly, methods for validating the TPHCWG approach.

5.1 FRACTION-SPECIFIC PROPERTIES IN THE LABORATORY

Laboratory studies can be designed to validate the values determined in Table 8 which define the physical-chemical properties of the fate and transport fractions. Briefly, a chemical mixture (e.g., gasoline) is added to vials containing distilled water. The mixture is allowed to equilibrate with the water over time. Once a steady-state has been reached, concentrations of each fraction are measured in the aqueous phase. These concentrations are compared to the fraction-specific water solubilities. Then the aqueous phase is introduced to vials containing headspace, and further partitioning of the fractions between air and water are measured. This partitioning is a measure of the Henry's law constant of the specific fraction. Vapor pressures can be measured directly or may be calculated and validated by multiplying the fraction-specific Henry's constant by the fraction-specific water solubility. For hydrocarbon mixtures (e.g., diesel, weathered products) analyses can be compared using Raoult's law.

5.2 LEACHING FACTORS AND VOLATILIZATION FACTORS IN THE LABORATORY

Laboratory studies can also be used to validate the multiphase partitioning of the chemical fractions. These studies can be done on clean soil that is spiked with a known quantity of a mixture or using soil from a contaminated site. The benefits of using clean soil from a representative area is that an exact mass balance on the fate and transport of the chemicals in the mixture can be completed. Losses due to volatilization, biodegradation, and sorption can be readily quantified. However, matching site conditions in the laboratory can be difficult, especially when dealing with variably weathered products.

Removing contaminated soil from a site and performing leach tests represents another option. The EPA has developed the Synthetic Precipitation Leaching Procedure (SPLP) to model an acid rain leaching environment. The SPLP is one potential leaching method for evaluating TPH contaminated sites. Other site-specific modifications of soil leaching tests could also be used. A benefit of this option is that the risk assessor does not need to rely on the partitioning equations for estimating leaching to groundwater. Also, the samples are directly applicable to the site in question (no extrapolation from a clean area soil is needed). However, if the initial concentrations in the soil cores are not characterized appropriately, the mass balance would be inaccurate and would lead to either an underestimate or overestimate of leaching. Also, providing and supporting the selection of samples as representative is necessary. Finally, depending on the types of soil and waste at the site, leaching tests that involve removing and filtering of samples could be problematic if the site contains very oily or sticky soils. For sites where an acid rain leachate is not applicable, site groundwater (or water designed to represent site groundwater) is an appropriate leachate.

5.3 FIELD VALIDATION OF FRACTION-SPECIFIC FATE AND TRANSPORT PARAMETERS

Site-specific data can also be used to validate the estimates of partitioning derived from the Working Group approach for fate and transport. As the leaching factor is a ratio of the fraction concentration in water to the fraction concentration in soil, direct calculations of LF based on site soil and groundwater concentrations can be made. Additionally, a modification of the volatilization factor (VF) can be used to compare concentrations in soil gas to concentrations in soil. The measured LF and VF values are then compared to modeled estimates using site-specific parameters in conjunction with fraction-specific chemical properties. Differences between measured and estimated values will highlight research needs and identify data gaps.

Factors for comparing site values of leaching and volatilization with expected estimates based on mathematical models include site-specific properties such as those described in Table 5. Table 9 provides a list of methods that could be used to gather the data. An important recommendation is that samples are not only taken spatially across the site but also at different depths. This will ensure that the site is adequately characterized.

In addition to the site properties, soil, soil vapor, and groundwater could be analyzed for the concentrations of the fate and transport fractions (again at different depths). Sampling requirements vary for the media; however, there are several techniques for analyzing many samples without drilling monitoring wells (e.g., Geoprobe,® Hydropunch). A Geoprobe can be used to collect soil vapor samples, soil samples, and groundwater samples. A Hydropunch is useful in collecting groundwater and NAPL samples. These simple techniques have some limitations. Field environmental professionals are the best source as to the applicability of each technique for any specific site.

To summarize, site-specific leaching and volatilization factors can be estimated using measured site physical parameters and the fraction-specific chemical properties or measured based on the comparison of fraction concentration in different media.

Table 9. Site Parameters for Validating Fraction-Specific Leaching Factors and Volatilization Factors (modified from U.S. EPA, 1996)

Parameter	Data Source	Method	Used to Calculate:
Dimensions of site (W, L_s)	Sampling data	Use sampling data to define the length, width, and depth of contaminated region	LF, VF
Soil texture	Lab measurement	Particle size analysis and USDA classification	θ_{ws}, I, other generic parameters
Dry soil bulk density (ρ_s)	Field measurement	ASTM D 2937 w/ ASTM 2216	LF, VF
Soil porosity (θ_t)	Equation	$\theta_t = 1 - \rho_s/\rho_p$ (where $\rho_p = 2.65$)	D_s^{eff}
Soil moisture content (θ_{ws})	Equation	Clapp and Hornberger (1978)	LF, VF, C_{sat}
Soil air content (θ_{as})	Equation	$\theta_{as} = \theta_t - \theta_{ws}$	LF. VF
Soil organic carbon content (f_{oc})	Lab measurement	Nelson and Sommers (1982)	k_s, C_{sat}
Mean annual wind speed (U_{air})	National Weather Service	Dataset of wind data; mean annual values of closest location in dataset	VF
Infiltration/recharge (I)	Regional estimate/ estimation models	Local knowledge or HELP model	LF
Hydraulic conductivity (K)	Field measurements, regional estimates	Aquifer tests, local knowledge	U_{gw}
Hydraulic gradient (i)	Field measurements, regional estimates	Site water level data, local knowledge	U_{gw}
Darcy velocity (U_{gw})	Equation	$U_{gw} = -K\,i$	LF

References

API (1992). *Technical Data Book-Petroleum Refining*, Fifth Edition, American Petroleum Institute, Washington, DC.

API (1993). *Petroleum Product Surveys*, American Petroleum Institute, Washington, DC.

ASTM (1995). *Standard Guide for Risk-Based Corrective Action Applied at Petroleum Release Sites.* Designation: E 1739-95. American Society for Testing and Materials, West Conshohocken, PA.

Bischoff, K.B., A. Nigam, and M.T. Klein (1991). "Lumping of discrete kinetic systems" in G. Astarita and S. I. Sandler (eds.). _*Kinetic and Thermodynamic Lumping of Multicomponent Mixtures*. Elsevier Science Publishers, Amsterdam, The Netherlands, pp. 33-48.

BP (1996). Summary tables of laboratory analysis for diesel and fuel oil #2, personal communication from B. Albertson, Friedman and Bruya, Inc., Seattle, WA, developed for British Petroleum.

Clapp, R.B. and G.M. Hornberger (1978). "Empirical equations for some soil hydraulic properties." *Water Resources Research* 14:601-604.

DiToro, D.M. (1985). "A particle interaction model of reversible organic chemical sorption." *Chemosphere* 14(10):1503-1538.

Eastcott, L., W.Y. Shiu, and D. Mackay (1988). "Environmentally relevant physical-chemical properties of hydrocarbons: A review of data and development of simple correlations." *Oil and Chemical Pollution* 4:191-216.

Feenstra, S., D.M. Mackay, and J.A. Cherry (1991). "A method for assessing residual NAPL based on organic chemical concentrations in soil samples." *Groundwater Monitoring Review*, Spring, 1991, p128-135.

GSC (1990). *Riskpro Software and Users Guide*. General Science Corporation, Laurel, MD.

Hillel, D. (1980). *Fundamentals of Soil Physics*. Academic Press, Orlando, FL.

Karickhoff, S.W., D.S. Brown, and T.A. Scott (1979). "Sorption of hydrophobic pollutants on natural sediments." *Water Research* 13:241-248.

Leo, A. and C. Hansch (1991). ClogP Program for Estimation of Octanol-Water Partition Coefficients, DAYLIGHT Chemical Information Systems, Inc., Irvine, CA.

LUFT (1988). *Leaking Underground Fuel Tank Manual: Guidelines for Site Assessment, Cleanup, and Underground Storage Tank Closure.* State of California Leaking Underground Fuel Tank Force, May 1988.

Lyman, W.J., W.F. Reehl, and D.H. Rosenblatt (1990). *Handbook of Chemical Property Estimation Methods: Environmental Behavior of Organic Compounds*. McGraw Hill Book Company, New York.

Lyman, W.J., P.J., Reidy, and B. Levy (1992). *"Contaminants Dissolved in Groundwater," in Mobility and Degradation of Organic Contaminants in Subsurface Environments.* C.K. Smoley, Inc., Chelsea, MI. p207-235.

Mackay, D., W.Y. Shui, and K.C. Ma (1993). *Illustrated Handbook of Physical-Chemical Properties and Environmental Fate for Organic Chemicals* Vols. I- IV. Lewis Publishers, Chelsea, MI.

McCabe, W.L., J.C. Smith, and P. Harriot (1985). *Unit Operations of Chemical Engineering,* 4th ed., McGraw-Hill Book Company, New York.

Millington, R.J. and J.P. Quirk (1961). "Permeability of porous solids," *Trans. Faraday Society* 57:1200-1207.

Nelson, D.W. and L.E. Sommers (1982). "Total carbon, organic carbon, and organic matter," in A.L. Page, ed. *Methods of Soil Analysis. Part 2. Chemical and Microbiological Properties. 2nd Edition* 9(2):539-579, American Society of Agronomy, Madison, WI.

Olsen, R.L. and A. Davis (1990). "Predicting the fate and transport of organic compounds in groundwater, Part 1" *Hazardous Materials Control* 3(3):38-63.

Peterson, D. (1994). "Calculating the aquatic toxicity of hydrocarbon mixtures." *Chemosphere,* 29(12):2493-2506.

Reid, R.C., J.M. Prausnitz, and T.K. Sherwood (1977). *Properties of Gases and Liquids. 3rd ed.,.* McGraw-Hill Book Company, New York.

Russom, C.L., E.B. Anderson, B.E. Greenwood, and A. Pilli. (1991) "ASTER: An integration of the AQUIRE data base and the QSAR system for use in ecological risk assessments." *The Science of the Total Environment* 109/110:667-670.

Sangster, J. (1989). "Octanol-water partition coefficients of simple organic compounds," *J. Physical and Chemical Reference Data* 18(3):1111-1229.

Smith, J.M. and H.C. Van Ness. (1975). *Introduction to Chemical Engineering Thermodynamics. 3rd ed.,* McGraw-Hill Book Company, New York.

U.S. EPA (1992). *Estimating Potential for Occurrence of DNAPL at Superfund Sites.* U.S. Environmental Protection Agency, Publication: 9355.4-07FS, Washington, DC, January 1992.

U.S. EPA (1996). *Soil Screening Guidance: User's Guide.* U.S. Environmental Protection Agency, Publication 9355.4-23, Washington, DC, April 1996.

Voice, T.C. and W.J. Weber (1983) "Sorption of hydrophobic compounds by sediments, soil and suspended solids." *Water Research* 10:1433-1441.

APPENDIX A

Petroleum Hydrocarbon Composition of Fuels Sorted by Fuel Type

Compound	Number of Carbons	EC	Weight Percent	Reference
CRUDE OIL				
Straight Chain Alkanes				
n-Hexane	6	6	0.7 - 1.8	API, 1993
n-Heptane	7	7	0.8 - 2.3	API, 1993
n-Octane	8	8	0.9 - 1.9	API, 1993
n-Nonane	9	9	0.6 - 1.9	API, 1993
n-Decane	10	10	1.8	API, 1993
n-Undecane	11	11	1.7	API, 1993
n-Dodecane	12	12	1.7	API, 1993
Branched Chain Alkanes				
2,2-Dimethylbutane	6	5.37	0.04	API, 1993
2,3-Dimethylbutane	6	5.68	0.04 - 0.14	API, 1993
2-Methylpentane	6	5.72	0.3 - 0.4	API, 1993
3-Methylpentane	6	5.85	0.3 - 0.4	API, 1993
3-Ethylpentane	7		0.05	API, 1993
2,4-Dimethylpentane	7	6.31	0.05	API, 1993
2,3-Dimethylpentane	7	6.69	0.1 - 0.6	API, 1993
2,2,4-Trimethylpentane	8	6.89	0.004	API, 1993
2,3,3-Trimethylpentane	8	7.58	0.006	API, 1993
2,3,4-Trimethylpentane	8	7.55	0.005	API, 1993
2-Methyl-3-ethylpentane	8	7.66	0.04	API, 1993
2-Methylhexane	7	6.68	0.7	API, 1993
3-Methylhexane	7	6.76	0.19 - 0.5	API, 1993
2,2-Dimethylhexane	8	7.25	0.01 - 0.1	API, 1993
2,3-Dimethylhexane	8	7.65	0.06 - 0.16	API, 1993
2,4-Dimethylhexane	8	7.38	0.06	API, 1993
2,5-Dimethylhexane	8	7.36	0.06	API, 1993
3,3-Dimethylhexane	8	7.45	0.03	API, 1993
2,3-Dimethylheptane	9	8.64	0.05	API, 1993
2,6-Dimethylheptane	9	8.47	0.05 -0.25	API, 1993
2-Methyloctane	9		0.4	API, 1993
3-Methyloctane	9	8.78	0.1 - 0.4	API, 1993
4-Methyloctane	9	8.71	0.1	API, 1993
Cycloalkanes				
Cyclopentane	5	5.66	0.05	API, 1993
Methylcyclopentane	6	6.27	0.3 - 0.9	API, 1993
1,1-Dimethylcyclopentane	7	6.72	0.06 - 0.2	API, 1993
1-trans-2-Dimethylcyclopentane	7	6.87	0.15 - .5	API, 1993
1-cis-3-Dimethylcyclopentane	7	6.82	0.2	API, 1993
1-trans-3-Dimethylcyclopentane	7	6.85	0.2 - 0.9	API, 1993
1,1,2-Trimethylcyclopentane	8	7.67	0.06	API, 1993
1,1,3-Trimethylcyclopentane	8	7.25	0.3	API, 1993
1-trans-2-cis-3-Trimethylcyclopentane	8	7.51	0.3 - 0.4	API, 1993

Compound	Number of Carbons	EC	Weight Percent	Reference
1-trans-2-cis-4-Trimethylcyclopentane	8		0.2	API, 1993
1-trans-2-Dimethylcyclohexane	8	7.94	0.3	API, 1993
Ethylcyclohexane	8	8.38	0.2	API, 1993
Cyclohexane	6	6.59	0.7	API, 1993
1-trans-2-trans-4-Trimethylcyclohexane	9		0.2	API, 1993
Alkyl Benzenes				
Benzene	6	6.5	0.04 - 0.4	API, 1993
Toluene	7	7.58	0.09 - 2.5	API, 1993
Ethylbenzene	8	8.5	0.09 - 0.31	API, 1993
o-Xylene	8	8.81	0.03 - 0.68	API, 1993
m-Xylene	8	8.6	0.08 - 2.0	API, 1993
p-Xylene	8	8.61	0.09 - 0.68	API, 1993
1-Methyl-4-ethylbenzene	9	9.57	0.03 - 0.13	API, 1993
1-Methyl-2-ethylbenzene	9	9.71	0.01 - 0.09	API, 1993
1-Methyl-3-ethylbenzene	9	9.55	0.04 - 0.4	API, 1993
1,2,3-Trimethylbenzene	9	10.06	0.1	API, 1993
1,2,4-Trimethylbenzene	9	9.84	0.13 - 0.69	API, 1993
1,3,5-Trimethylbenzene	9	9.62	0.05 - 0.18	API, 1993
1,2,3,4-Tetramethylbenzene	10	11.57	0.2	API, 1993
Biphenyl	12	14.26	0.006 - .04	API, 1993
Naphtheno-Benzenes				
Indan	9	10.27	0.07	API, 1993
Tetralin (tetrahydronaphthalene)	10	11.7	0.03	API, 1993
5-Methylthtrohydronaphthalene	11		0.08	API, 1993
6-Methylthtrohydronaphthalene	11		0.09	API, 1993
Fluorene	13	16.55	0.003 - 0.06	API, 1993
Alkyl Naphthalenes				
Naphthalene	10	11.69	0.02 - 0.09	API, 1993
Polynuclear Aromatics				
Phenanthrene	14	19.36	0.003 - 0.05	API, 1993
DIESEL				
Straight Chain Alkanes				
n-Octane	8	8	0.1	BP, 1996
n-Nonane	9	9	0.19 - 0.49	BP, 1996
n-Decane	10	10	0.28 - 1.2	BP, 1996
n-Undecane	11	11	0.57 - 2.3	BP, 1996
n-Dodecane	12	12	1.0 - 2.5	BP, 1996
n-Tridecane	13	13	1.5 - 2.8	BP, 1996
n-Tetradecane	14	14	0.61 - 2.7	BP, 1996
n-Pentadecane	15	15	1.9 - 3.1	BP, 1996
n-Hexadecane	16	16	1.5 - 2.8	BP, 1996
n-Heptadecane	17	17	1.4 - 2.9	BP, 1996

Compound	Number of Carbons	EC	Weight Percent	Reference
n-Octadecane	18	18	1.2 - 2.0	BP, 1996
n-Nonadecane	19	19	0.7 - 1.5	BP, 1996
n-Eicosane	20	20	0.4 - 1.0	BP, 1996
n-Heneicosane	21	21	0.26 - 0.83	BP, 1996
n-Docosane	22	22	0.14 - 0.44	BP, 1996
n-Tetracosane	24	24	0.35	BP, 1996
Branched Chain Alkanes				
3-Methylundecane	12		0.09 - 0.28	BP, 1996
2-Methyldodecane	13		0.15 - 0.52	BP, 1996
3-Methyltridecane	14		0.13 - 0.30	BP, 1996
2-Methyltetradecane	15		0.34 - 0.63	BP, 1996
Alkyl Benzenes				
Benzene	6	6.5	0.003 - 0.10	BP, 1996
Toluene	7	7.58	0.007 - 0.70	BP, 1996
Ethylbenzene	8	8.5	0.007 - 0.20	BP, 1996
o-Xylene	8	8.81	.001 - 0.085	BP, 1996
m-Xylene	8	8.6	0.018 - 0.512	BP, 1996
p-Xylene	8	8.61	0.018 - 0.512	BP, 1996
Styrene	9	8.83	<.002	BP, 1996
1-Methyl-4-isopropylbenzene	10	10.13	0.003 - 0.026	BP, 1996
1,3,5-Trimethylbenzene	9	9.62	0.09 - 0.24	BP, 1996
n-Propylbenzene	9	9.47	0.03 - 0.048	BP, 1996
Isopropylbenzene	9	9.13	<0.01	BP, 1996
n-Butylbenzene	10	10.5	0.031 - 0.046	BP, 1996
Biphenyl	12		0.01 - 0.12	BP, 1996
Naphtheno-Benzenes				
Fluorene	13	16.55	0.034 - 0.15	BP, 1996
Fluoranthene	16	21.85	0.0000007 - 0.02	BP, 1996
Benz(b)fluoranthene	20	30.14	0.0000003 - 0.000194	BP, 1996
Benz(k)fluoranthene	20	30.14	0.0000003 - 0.000195	BP, 1996
Indeno (1,2,3-cd) pyrene	22	35.01	0.000001 - 0.000097	BP, 1996
Alkyl Naphthalenes				
Naphthalene	10	11.69	.01 - 0.80	BP, 1996
1-Methylnaphthalene	11	12.99	0.001 - 0.81	BP, 1996
2-Methylnaphthalene	11	12.84	0.001 - 1.49	BP, 1996
1,3-Dimethylnaphthalene	12	14.77	0.55 - 1.28	BP, 1996
1,4-Dimethylnaphthalene	12	14.6	0.110 - 0.23	BP, 1996
1,5-Dimethylnaphthalene	12	13.87	0.16 - 0.36	BP, 1996
Polynuclear Aromatics				
Anthracene	14	19.43	0.000003 -0.02	BP, 1996
2-Methyl anthracene	15	20.73	0.000015 - 0.018	BP, 1996
Phenanthrene	14	19.36	0.000027 - 0.30	BP, 1996
1-Methylphenanthrene	15	20.73	0.000011 - 0.024	BP, 1996
2-Methylphenanthrene	15		0.014 - 0.18	BP, 1996

Compound	Number of Carbons	EC	Weight Percent	Reference
3-Methylphenanthrene	15		0.000013 - 0.011	BP, 1996
4 & 9-Methylphenanthrene	15		0.00001 - 0.034	BP, 1996
Pyrene	16	20.8	0.000018 - 0.015	BP, 1996
1-Methylpyrene	17		0.0000024 - 0.00137	BP, 1996
2-Methylpyrene	17		0.0000037 - 0.00106	BP, 1996
Benz(a)anthracene	18	26.37	0.0000021 - 0.00067	BP, 1996
Chrysene	18	27.41	0.000045	BP, 1996
Triphenylene	18	26.61	0.00033	BP, 1996
Cyclopenta(cd)pyrene	18		0.000002 - 0.0000365	BP, 1996
1-Methyl-7-isopropylphenanthrene	18		0.0000015 - 0.00399	BP, 1996
3-Methylchrysene	19		<0.001	BP, 1996
6-Methylchrysene	19		<0.0005	BP, 1996
Benz(a)pyrene	20	31.34	0.000005 -0.00084	BP, 1996
Benz(e)pyrene	20	31.17	0.0000054 - 0.000240	BP, 1996
Perylene	20	31.34	<0.0001	BP, 1996
Benz(ghi)perylene	22	34.01	0.0000009 - 0.00004	BP, 1996
Picene	22		0.0000004 - 0.000083	BP, 1996
FUEL OIL #2				
Straight Chain Alkanes				
n-Octane	8	8	0.1	BP, 1996
n-Nonane	9	9	0.20 - 0.30	BP, 1996
n-Decane	10	10	0.5	BP, 1996
n-Undecane	11	11	0.80 - 0.90	BP, 1996
n-Dodecane	12	12	0.84 - 1.20	BP, 1996
n-Tridecane	13	13	0.96 - 2.00	BP, 1996
n-Tetradecane	14	14	1.03 - 2.50	BP, 1996
n-Pentadecane	15	15	1.13 - 3.20	BP, 1996
n-Hexadecane	16	16	1.05 - 3.30	BP, 1996
n-Heptadecane	17	17	0.65 - 3.60	BP, 1996
n-Octadecane	18	18	0.55 - 2.50	BP, 1996
n-Nonadecane	19	19	0.33 - 1.30	BP, 1996
n-Eicosane	20	20	0.18 - 0.60	BP, 1996
n-Heneicosane	21	21	0.09 - 0.40	BP, 1996
n-Docosane	22	22	0.1	BP, 1996
Alkyl Benzenes				
Benzene	6	6.5	<0.125	BP, 1996
Toluene	7	7.58	0.025 - 0.110	BP, 1996
Ethylbenzene	8	8.5	0.028 - 0.04	BP, 1996
Biphenyl	12		0.006 - 0.009	BP, 1996
Naphtheno-Benzenes				
Acenaphthene	12	15.5	0.013 - 0.022	BP, 1996
Acenaphthylene	12	15.06	0.006	BP, 1996
Fluorene	13	16.55	0.004 - 0.045	BP, 1996

Compound	Number of Carbons	EC	Weight Percent	Reference
Fluoranthene	16	21.85	0.000047 - 0.00037	BP, 1996
2,3- Benzofluorene	17	23.83	<0.0024	BP, 1996
Benzo(a)fluorene	17		<0.0006	BP, 1996
Benzo(ghi)fluoranthene	18		<0.0024	BP, 1996
Benz(b)fluoranthene	20	30.14	<0.0024	BP, 1996
Benz(k)fluoranthene	20	30.14	<0.00006	BP, 1996
Indeno (1,2,3-cd) pyrene	22	35.01	<0.0012	BP, 1996
Alkyl Naphthalenes				
Naphthalene	10	11.69	0.009 - 0.40	BP, 1996
1-Methylnaphthalene	11	12.99	0.29 - 0.48	BP, 1996
2-Methylnaphthalene	11	12.84	0.36 -1.00	BP, 1996
1,4-Dimethylnaphthalene	12	14.6	0.043 - 0.045	BP, 1996
Polynuclear Aromatics				
Anthracene	14	19.43	0.00010 - 0.011	BP, 1996
2-Methyl anthracene	15	20.73	0.009 - 0.017	BP, 1996
9,10-Dimethyl anthracene	16		0.002 - 0.006	BP, 1996
Phenanthrene	14	19.36	0.009 -0.170	BP, 1996
1-Methylphenanthrene	15	20.73	0.017	BP, 1996
2-Methylphenanthrene	15		0.768	BP, 1996
Pyrene	16	20.8	0.00 - 0.012	BP, 1996
Benz(a)anthracene	18	26.37	0.000002 - 0.00012	BP, 1996
Chrysene	18	27.41	0.000037 - 0.00039	BP, 1996
Triphenylene	18	26.61	0.00002 - 0.00014	BP, 1996
Benzo(b)chrysene	19		<0.0036	BP, 1996
Benz(a)pyrene	20	31.34	0.000001 - 0.000060	BP, 1996
Benz(e)pyrene	20	31.17	0.0000020 - 0.000010	BP, 1996
Benzo(ghi)pyrene	20	31.17	0.0000010 - 0.0000070	BP, 1996
Perylene	20	31.34	<0.0024	BP, 1996
3-Methylcholanthrene	21		<0.00006	BP, 1996
Benz(ghi)perylene	22	34.01	0.0000057	BP, 1996
Picene	22		<0.00012	BP, 1996
Coronene	24	34.01	<0.000024	BP, 1996
GASOLINE				
Straight Chain Alkanes				
Propane	3	3	0.01 - 0.14	LUFT, 1988
n-Butane	4	4	3.93 - 4.70	LUFT, 1988
n-Pentane	5	5	5.75 - 10.92	LUFT, 1988
n-Hexane	6	6	0.24 - 3.50	LUFT, 1988
n-Heptane	7	7	0.31 - 1.96	LUFT, 1988
n-Octane	8	8	0.36 - 1.43	LUFT, 1988
n-Nonane	9	9	0.07 - 0.83	LUFT, 1988
n-Decane	10	10	0.04 - 0.50	LUFT, 1988
n-Undecane	11	11	0.05 - 0.22	LUFT, 1988
n-Dodecane	12	12	0.04 - 0.09	LUFT, 1988

Compound	Number of Carbons	EC	Weight Percent	Reference
Branched Chain Alkanes				
Isobutane	4	3.67	0.12 - 0.37	LUFT, 1988
2,2-Dimethylbutane	6	5.37	0.17 - 0.84	LUFT, 1988
2,3-Dimethylbutane	6	5.68	0.59 - 1.55	LUFT, 1988
2,2,3-Trimethylbutane	7	6.36	0.01 - 0.04	LUFT, 1988
Neopentane	5	4.32	0.02 - 0.05	LUFT, 1988
Isopentane	5	4.75	6.07 - 10.17	LUFT, 1988
2-Methylpentane	6	5.72	2.91 - 3.85	LUFT, 1988
3-Methylpentane	6	5.85	2.4 (vol)	LUFT, 1988
2,4-Dimethylpentane	7	6.31	0.23 - 1.71	LUFT, 1988
2,3-Dimethylpentane	7	6.69	0.32 - 4.17	LUFT, 1988
3,3-Dimethylpentane	7	6.55	0.02 - 0.03	LUFT, 1988
2,2,3-Trimethylpentane	8	7.37	0.09 - 0.23	LUFT, 1988
2,2,4-Trimethylpentane	8	6.89	0.32 - 4.58	LUFT, 1988
2,3,3-Trimethylpentane	8	7.58	0.05 - 2.28	LUFT, 1988
2,3,4-Trimethylpentane	8	7.55	0.11 - 2.80	LUFT, 1988
2,4-Dimethyl-3-ethylpentane	9		0.03 - 0.07	LUFT, 1988
2-Methylhexane	7	6.68	0.36 - 1.48	LUFT, 1988
3-Methylhexane	7	6.76	0.30 - 1.77	LUFT, 1988
2,4-Dimethylhexane	8	7.38	0.34 - 0.82	LUFT, 1988
2,5-Dimethylhexane	8	7.36	0.24 - 0.52	LUFT, 1988
3,4-Dimethylhexane	8	7.74	0.16 - 0.37	LUFT, 1988
3-Ethylhexane	8	7.79	0.01	LUFT, 1988
2-Methyl-3-ethylhexane	9		0.04 - 0.13	LUFT, 1988
2,2,4-Trimethylhexane	9	7.93	0.11 - 0.18	LUFT, 1988
2,2,5-Trimethylhexane	9	7.87	0.17 - 5.89	LUFT, 1988
2,3,3-Trimethylhexane	9		0.05 - 0.12	LUFT, 1988
2,3,5-Trimethylhexane	9	8.24	0.05 - 1.09	LUFT, 1988
2,4,4-Trimethylhexane	9	8.07	0.02 - 0.16	LUFT, 1988
2-Methylheptane	8	7.71	0.48 - 1.05	LUFT, 1988
3-Methylheptane	8	7.78	0.63 - 1.54	LUFT, 1988
4-Methylheptane	8	7.72	0.22 - 0.52	LUFT, 1988
2,2-Dimethylheptane	9	8.28	0.01 - 0.08	LUFT, 1988
2,3-Dimethylheptane	9	8.64	0.13 - 0.51	LUFT, 1988
2,6-Dimethylheptane	9	8.47	0.07 - 0.23	LUFT, 1988
3,3-Dimethylheptane	9	8.42	0.01 - 0.08	LUFT, 1988
3,4-Dimethylheptane	9	8.62	0.07 - 0.33	LUFT, 1988
2,2,4-Trimethylheptane	10		0.12 - 1.70	LUFT, 1988
3,3,5-Trimethylheptane	10		0.02 - 0.06	LUFT, 1988
3-Ethylheptane	9	8.77	0.02 - 0.16	LUFT, 1988
2-Methyloctane	9		0.14 - 0.62	LUFT, 1988
3-Methyloctane	9	8.78	0.34 - 0.85	LUFT, 1988
4-Methyloctane	9	8.71	0.11 - 0.55	LUFT, 1988
2,6-Dimethyloctane	10	9.32	0.06 - 0.12	LUFT, 1988
2-Methylnonane	10	9.72	0.06 - 0.41	LUFT, 1988

Compound	Number of Carbons	EC	Weight Percent	Reference
3-Methylnonane	10	9.78	0.06 - 0.32	LUFT, 1988
4-Methylnonane	10		0.04 - 0.26	LUFT, 1988
Cycloalkanes				
Cyclopentane	5	5.66	0.19 - 0.58	LUFT, 1988
Methylcyclopentane	6	6.27	not quantified	LUFT, 1988
1-Methyl-cis-2-ethylcyclopentane	8		0.06 - 0.11	LUFT, 1988
1-Methyl-trans-3-ethylcyclopentane	8		0.06 - 0.12	LUFT, 1988
1-cis-2-Dimethylcyclopentane	7	7.21	0.07 - 0.13	LUFT, 1988
1-trans-2-Dimethylcyclopentane	7	6.87	0.06 - 0.20	LUFT, 1988
1,1,2-Trimethylcyclopentane	8	7.67	0.06 - 0.11	LUFT, 1988
1-trans-2-cis-3-Trimethylcyclopentane	8	7.51	0.01 - 0.25	LUFT, 1988
1-trans-2-cis-4-Trimethylcyclopentane	8		0.03 - 0.16	LUFT, 1988
Ethylcyclopentane	7	7.34	0.14 - 0.21	LUFT, 1988
n-Propylcyclopentane	8	7.1	0.01 - 0.06	LUFT, 1988
Isopropylcyclopentane	8		0.01 - 0.02	LUFT, 1988
1-trans-3-Dimethylcyclohexane	8	7.99	0.05 - 0.12	LUFT, 1988
Ethylcyclohexane	8	8.38	0.17 - 0.42	LUFT, 1988
Cyclohexane	6	6.59	0.08	API, 1993
Straight Chained Alkenes				
cis-2-Butene	4	4.25	0.13 - 0.17	LUFT, 1988
trans-2-Butene	4	4.1	0.16 - 0.20	LUFT, 1988
Pentene-1	5	4.89	0.33 - 0.45	LUFT, 1988
cis-2-Pentene	5	5.16	0.43 - 0.67	LUFT, 1988
trans-2-Pentene	5	5.08	0.52 - 0.90	LUFT, 1988
cis-2-Hexene	6	6.14	0.15 - 0.24	LUFT, 1988
trans-2-Hexene	6	6.05	0.18 - 0.36	LUFT, 1988
cis-3-Hexene	6	6.03	0.11 - 0.13	LUFT, 1988
trans-3-Hexene	6	6.02	0.12 - 0.15	LUFT, 1988
cis-3-Heptene	7	7.01	0.14 - 0.17	LUFT, 1988
trans-2-Heptene	7	7.05	0.06 - 0.10	LUFT, 1988
Branched Chain Alkenes				
2-Methyl-1-butene	5	4.96	0.22 - 0.66	LUFT, 1988
3-Methyl-1-butene	5	4.57	0.08 - 0.12	LUFT, 1988
2-Methyl-2-butene	5	5.21	0.96 - 1.28	LUFT, 1988
2,3-Dimethyl-1-butene	6	5.7	0.08 - 0.10	LUFT, 1988
2-Methyl-1-pentene	6	5.89	0.20 - 0.22	LUFT, 1988
2,3-Dimethyl-1-pentene	7		0.01 - 0.02	LUFT, 1988
2,4-Dimethyl-1-pentene	7	6.48	0.02 - 0.03	LUFT, 1988
4,4-Dimethyl-1-pentene	7		0.60 (vol)	LUFT, 1988
2-Methyl-2-pentene	6	6.07	0.27 - 0.32	LUFT, 1988
3-Methyl-cis-2-pentene	6	6.11	0.35 - 0.45	LUFT, 1988
3-Methyl-trans-2-pentene	6	6.22	0.32 - 0.44	LUFT, 1988
4-Methyl-cis-2-pentene	6	5.69	0.04 - 0.05	LUFT, 1988

Compound	Number of Carbons	EC	Weight Percent	Reference
4-Methyl-trans-2-pentene	6	5.73	0.08 - 0.30	LUFT, 1988
4,4-Dimethyl-cis-2-pentene	7	6.47	0.02	LUFT, 1988
4,4-Dimethyl-trans-2-pentene	7	6.23	Not quantified	LUFT, 1988
3-Ethyl-2-pentene	7	7.07	0.03 - 0.04	LUFT, 1988
Cycloalkenes				
Cyclopentene	5	5.55	0.12 - 0.18	LUFT, 1988
3-Methylcyclopentene	6	6.1	0.03 - 0.08	LUFT, 1988
Cyclohexene	6	6.74	0.03	LUFT, 1988
Alkyl Benzenes				
Benzene	6	6.5	0.12 - 3.50	LUFT, 1988
Toluene	7	7.58	2.73 - 21.80	LUFT, 1988
Ethylbenzene	8	8.5	0.36 - 2.86	LUFT, 1988
o-Xylene	8	8.81	0.68 - 2.86	LUFT, 1988
m-Xylene	8	8.6	1.77 - 3.87	LUFT, 1988
p-Xylene	8	8.61	0.77 - 1.58	LUFT, 1988
1-Methyl-4-ethylbenzene	9	9.57	0.18 - 1.00	LUFT, 1988
1-Methyl-2-ethylbenzene	9	9.71	0.19 - 0.56	LUFT, 1988
1-Methyl-3-ethylbenzene	9	9.55	0.31 - 2.86	LUFT, 1988
1-Methyl-2-n-propylbenzene	10		0.01 - 0.17	LUFT, 1988
1-Methyl-3-n-propylbenzene	10		0.08 - 0.56	LUFT, 1988
1-Methyl-2-isopropylbenzene	10		0.01 - 0.12	LUFT, 1988
1-Methyl-3-t-butylbenzene	11		0.03 - 0.11	LUFT, 1988
1-Methyl-4-t-butylbenzene	11	10.92	0.04 - 0.13	LUFT, 1988
1,2-Dimethyl-3-ethylbenzene	10	10.93	0.02 - 0.19	LUFT, 1988
1,2-Dimethyl-4-ethylbenzene	10	10.75	0.50 - 0.73	LUFT, 1988
1,3-Dimethyl-2-ethylbenzene	10	10.81	0.21 - 0.59	LUFT, 1988
1,3-Dimethyl-4-ethylbenzene	10	10.75	0.03 - 0.44	LUFT, 1988
1,3-Dimethyl-5-ethylbenzene	10	10.51	0.11 - 0.42	LUFT, 1988
1,3-Dimethyl-5-t-butylbenzene	12		0.02 - 0.16	LUFT, 1988
1,4-Dimethyl-2-ethylbenzene	10	10.68	0.05 - 0.36	LUFT, 1988
1,2,3-Trimethylbenzene	9	10.06	0.21 - 0.48	LUFT, 1988
1,2,4-Trimethylbenzene	9	9.84	0.66 - 3.30	LUFT, 1988
1,3,5-Trimethylbenzene	9	9.62	0.13 - 1.15	LUFT, 1988
1,2,3,4-Tetramethylbenzene	10	11.57	0.02 - 0.19	LUFT, 1988
1,2,3,5-Tetramethylbenzene	10	11.09	0.14 - 1.06	LUFT, 1988
1,2,4,5-Tetramethylbenzene	10	11.05	0.05 - 0.67	LUFT, 1988
1,2-Diethylbenzene	10	10.52	0.57	LUFT, 1988
1,3-Diethylbenzene	10	10.4	0.05 - 0.38	LUFT, 1988
n-Propylbenzene	9	9.47	0.08 - 0.72	LUFT, 1988
Isopropylbenzene	9	9.13	<10.01 - 0.23	LUFT, 1988
n-Butylbenzene	10	10.5	0.04 - 0.44	LUFT, 1988
Isobutylbenzene	10	9.96	0.01 - 0.08	LUFT, 1988
sec-Butylbenzene	10	9.98	0.01 - 0.13	LUFT, 1988
t-Butylbenzene	10	9.84	0.12	LUFT, 1988
n-Pentylbenzene	11	11.49	0.01 - 0.14	LUFT, 1988
Isopentylbenzene	11		0.07 - 0.17	LUFT, 1988

Compound	Number of Carbons	EC	Weight Percent	Reference
Naphtheno Benzenes				
Indan	9	10.27	0.25 - 0.34	LUFT, 1988
1-Methylindan	10		0.04 - 0.17	LUFT, 1988
2-Methylindan	10	11.39	0.02 - 0.10	LUFT, 1988
4-Methylindan	10	11.33	0.01 - 0.16	LUFT, 1988
5-Methylindan	10	11.28	0.09 - 0.30	LUFT, 1988
Tetralin (tetrahydronaphthalene)	10	11.7	0.01 - 0.14	LUFT, 1988
Alkyl Naphthalenes				
Naphthalene	10	11.69	0.09 - 0.49	LUFT, 1988
Polynuclear Aromatics				
Pyrene	16	20.8	Not quantified	LUFT, 1988
Benz(a)anthracene	18	26.37	Not quantified	LUFT, 1988
Benz(a)pyrene	20	31.34	0.19 - 2.8 mg/kg	LUFT, 1988
Benz(e)pyrene	20	31.17	Not quantified	LUFT, 1988
Benz(ghi)perylene	22	34.01	Not quantified	LUFT, 1988
JP-4				
Straight Chain Alkanes				
n-Butane	4	4	0.12	API, 1993
n-Pentane	5	5	1.06	API, 1993
n-Hexane	6	6	2.21	API, 1993
n-Heptane	7	7	3.67	API, 1993
n-Octane	8	8	3.8	API, 1993
n-Nonane	9	9	2.25	API, 1993
n-Decane	10	10	2.16	API, 1993
n-Undecane	11	11	2.32	API, 1993
n-Dodecane	12	12	2	API, 1993
n-Tridecane	13	13	1.52	API, 1993
n-Tetradecane	14	14	0.73	API, 1993
Branched Chain Alkanes				
Isobutane	4	3.67	0.66	API, 1993
2,2-Dimethylbutane	6	5.37	0.1	API, 1993
2,2,3,3-Tetramethylbutane	8	7.3	0.24	API, 1993
2-Methylpentane	6	5.72	1.28	API, 1993
3-Methylpentane	6	5.85	0.89	API, 1993
2,2-Dimethylpentane	7	6.25	0.25	API, 1993
2-Methylhexane	7	6.68	2.35	API, 1993
3-Methylhexane	7	6.76	1.97	API, 1993
2,2-Dimethylhexane	8	7.25	0.71	API, 1993
2,4-Dimethylhexane	8	7.38	0.58	API, 1993
2,5-Dimethylhexane	8	7.36	0.37	API, 1993
3,3-Dimethylhexane	8	7.45	0.26	API, 1993
2-Methylheptane	8	7.71	2.7	API, 1993
3-Methylheptane	8	7.78	3.04	API, 1993

Compound	Number of Carbons	EC	Weight Percent	Reference
4-Methylheptane	8	7.72	0.92	API, 1993
2,4-Dimethylheptane	9	8.34	0.43	API, 1993
2,5-Dimethylheptane	9	8.47	0.52	API, 1993
4-Ethylheptane	9	8.69	0.18	API, 1993
2-Methyloctane	9		0.88	API, 1993
3-Methyloctane	9	8.78	0.79	API, 1993
4-Methyloctane	9	8.71	0.86	API, 1993
2-Methylundecane	12		0.64	API, 1993
2,6-Dimethylundecane	13		0.71	API, 1993
Cycloalkanes				
Methylcyclopentane	6	6.27	1.16	API, 1993
1-cis-2-Dimethylcyclopentane	7	7.21	0.54	API, 1993
1-cis-3-Dimethylcyclopentane	7	6.82	0.34	API, 1993
1-trans-3-Dimethylcyclopentane	7	6.85	0.36	API, 1993
Ethylcyclopentane	7	7.34	0.26	API, 1993
1-cis-3-Dimethylcyclohexane	8	7.75	0.42	API, 1993
Cyclohexane	6	6.59	1.24	API, 1993
Methylcyclohexane	7	7.22	2.27	API, 1993
1-Methyl-2-ethylcyclohexane	9		0.39	API, 1993
1-Methyl-3-ethylcyclohexane	9		0.17	API, 1993
1,3,5-Trimethylcyclohexane	9		0.99	API, 1993
1,1,3-Trimethylcyclohexane	9	8.45	0.48	API, 1993
n-Butylcyclohexane	10		0.7	API, 1993
Alkyl Benzenes				
Benzene	6	6.5	0.5	API, 1993
Toluene	7	7.58	1.33	API, 1993
Ethylbenzene	8	8.5	0.37	API, 1993
o-Xylene	8	8.81	1.01	API, 1993
m-Xylene	8	8.6	0.96	API, 1993
p-Xylene	8	8.61	0.35	API, 1993
1-Methyl-4-ethylbenzene	9	9.57	0.43	API, 1993
1-Methyl-2-ethylbenzene	9	9.71	0.23	API, 1993
1-Methyl-3-ethylbenzene	9	9.55	0.49	API, 1993
1-Methyl-2-isopropylbenzene	10		0.29	API, 1993
1,2-Dimethyl-4-ethylbenzene	10	10.75	0.77	API, 1993
1,3-Dimethyl-5-ethylbenzene	10	10.51	0.61	API, 1993
1,4-Dimethyl-2-ethylbenzene	10	10.68	0.7	API, 1993
1,2,4-Trimethylbenzene	9	9.84	1.01	API, 1993
1,3,5-Trimethylbenzene	9	9.62	0.42	API, 1993
1,3-Diethylbenzene	10	10.4	0.46	API, 1993
n-Propylbenzene	9	9.47	0.71	API, 1993
Isopropylbenzene	9	9.13	0.3	API, 1993

Compound	Number of Carbons	EC	Weight Percent	Reference
Alkyl Naphthalenes				
Naphthalene	10	11.69	0.5	API, 1993
1-Methylnaphthalene	11	12.99	0.78	API, 1993
2-Methylnaphthalene	11	12.84	0.56	API, 1993
2,6-Dimethylnaphthalene	12	14.6	0.25	API, 1993
JP-5				
Straight Chain Alkanes				
n-Octane	8	8	0.12	API, 1993
n-Nonane	9	9	0.38	API, 1993
n-Decane	10	10	1.79	API, 1993
n-Undecane	11	11	3.95	API, 1993
n-Dodecane	12	12	3.94	API, 1993
n-Tridecane	13	13	3.45	API, 1993
n-Tetradecane	14	14	2.72	API, 1993
n-Pentadecane	15	15	1.67	API, 1993
n-Hexadecane	16	16	1.07	API, 1993
n-Heptadecane	17	17	0.12	API, 1993
Branched Chain Alkanes				
2,4,6-Trimethylheptane	10		0.07	API, 1993
3-Methyloctane	9	8.78	0.07	API, 1993
4-Methyldecane	11		0.78	API, 1993
2-Methyldecane	11		0.61	API, 1993
2,6-Dimethyldecane	12		0.72	API, 1993
2-Methylundecane	12		1.39	API, 1993
2,6-Dimethylundecane	13		2	API, 1993
Cycloalkanes				
1-Methyl-4-ethylcyclohexane	9		0.48	API, 1993
1,3,5-Trimethylcyclohexane	9		0.09	API, 1993
1,1,3-Trimethylcyclohexane	9	8.45	0.05	API, 1993
n-Butylcyclohexane	10		0.9	API, 1993
Heptylcyclohexane	13		0.99	API, 1993
Straight Chain Alkenes				
Tridecene	13		0.45	API, 1993
Alkyl Benzenes				
o-Xylene	8	8.81	0.09	API, 1993
m-Xylene	8	8.6	0.13	API, 1993
1,2,4-Trimethylbenzene	9	9.84	0.37	API, 1993
1,3-Diethylbenzene	10	10.4	0.61	API, 1993
1,4-Diethylbenzene	10	10.46	0.77	API, 1993
1,2,4-Triethylbenzene	12	12.29	0.72	API, 1993
1-t-Butyl-3,4,5-trimethylbenzene	13		0.24	API, 1993

Compound	Number of Carbons	EC	Weight Percent	Reference
n-Heptylbenzene	13		0.27	API, 1993
n-Octylbenzene	14		0.78	API, 1993
Biphenyl	12		0.7	API, 1993
Phenylcyclohexane	12		0.82	API, 1993
Alkyl Naphthalenes				
Naphthalene	10	11.69	0.57	API, 1993
1-Methylnaphthalene	11	12.99	1.44	API, 1993
2-Methylnaphthalene	11	12.84	1.38	API, 1993
2,3-Dimethylnaphthalene	12	15	0.46	API, 1993
2,6-Dimethylnaphthalene	12	14.6	1.12	API, 1993
1-Ethylnaphthalene	12	14.41	0.32	API, 1993
JP-8				
Straight Chain Alkanes				
n-Heptane	7	7	0.03	API, 1993
n-Octane	8	8	0.9	API, 1993
n-Nonane	9	9	0.31	API, 1993
n-Decane	10	10	1.31	API, 1993
n-Undecane	11	11	4.13	API, 1993
n-Dodecane	12	12	4.72	API, 1993
n-Tridecane	13	13	4.43	API, 1993
n-Tetradecane	14	14	2.99	API, 1993
n-Pentadecane	15	15	1.61	API, 1993
n-Hexadecane	16	16	0.45	API, 1993
n-Heptadecane	17	17	0.08	API, 1993
n-Octadecane	18	18	0.02	API, 1993
Branched Chain Alkanes				
2,4,6-Trimethylheptane	10		0.07	API, 1993
3-Methyloctane	9	8.78	0.04	API, 1993
2-Methyldecane	11		0.41	API, 1993
2,6-Dimethyldecane	12		0.66	API, 1993
2-Methylundecane	12		1.16	API, 1993
2,6-Dimethylundecane	13		2.06	API, 1993
Cycloalkanes				
1-Methyl-4-ethylcyclohexane	9		0.1	API, 1993
1,3,5-Trimethylcyclohexane	9		0.06	API, 1993
1,1,3-Trimethylcyclohexane	9	8.45	0.06	API, 1993
n-Butylcyclohexane	10		0.74	API, 1993
n-Propylcyclohexane	9		0.14	API, 1993
Hexylcyclohexane	12		0.93	API, 1993
Heptylcyclohexane	13		1	API, 1993
Straight Chain Alkenes				
Tridecene	14		0.73	API, 1993

Compound	Number of Carbons	EC	Weight Percent	Reference
Alkyl Benzenes				
o-Xylene	8	8.81	0.06	API, 1993
m-Xylene	8	8.6	0.06	API, 1993
1-Methyl-2-isopropylbenzene	10		0.56	API, 1993
1,3-Dimethyl-5-ethylbenzene	10	10.51	0.62	API, 1993
1,2,4-Trimethylbenzene	9	9.84	0.27	API, 1993
1,2,4-Triethylbenzene	12	12.29	0.99	API, 1993
1,3,5-Triethylbenzene	12	12.1	0.6	API, 1993
n-Heptylbenzene	13		0.25	API, 1993
n-Octylbenzene	14		0.61	API, 1993
Biphenyl	12		0.63	API, 1993
Phenylcyclohexane	12		0.87	API, 1993
Alkyl Naphthalenes				
Naphthalene	10	11.69	1.14	API, 1993
1-Methylnaphthalene	11	12.99	1.84	API, 1993
2-Methylnaphthalene	11	12.84	1.46	API, 1993
2,3-Dimethylnaphthalene	12	15	0.36	API, 1993
2,6-Dimethylnaphthalene	12	14.6	1.34	API, 1993
1-Ethylnaphthalene	12	14.41	0.33	API, 1993
Straight Chain Alkanes				
n-Heptane	7	7	0.1	API, 1993
n-Octane	8	8	0.2 - 0.3	API, 1993
n-Nonane	9	9	0.4 - 0.8	API, 1993
n-Decane	10	10	1.5 - 1.7	API, 1993
n-Undecane	11	11	3.5 - 6.1	API, 1993
n-Dodecane	12	12	2.8 - 5.7	API, 1993
n-Tridecane	13	13	3.1 - 5.2	API, 1993
n-Tetradecane	14	14	2.3 - 4.7	API, 1993
n-Pentadecane	15	15	0.6 - 2.3	API, 1993
n-Hexadecane	16	16	0.1 - 0.7	API, 1993
n-Heptadecane	17	17	0.4	API, 1993
n-Octadecane	18	18	0.3	API, 1993
n-Nonadecane	19	19	0.2	API, 1993
n-Eicosane	20	20	0.1	API, 1993
n-Heneicosane	21	21	0.1	API, 1993

APPENDIX B

References for Physical-Chemical Properties

Compound	Molecular Weight	S mg/L	Specific Gravity	VP atm	H(c/c)	log K_{ow}	BP C	Dair cm²/s	Dwater cm²/s
Straight Chain Alkanes									
Propane	3	10	3	3	10	10	3	16	
n-Butane	9	9	9	9	9	10	9	5	
n-Pentane	9	9	9	9	9	9	9	5	
n-Hexane	9	9	9	9	9	9	9	12	12
n-Heptane	9	9	9	9	9	9	9	5	
n-Octane	9	9	9	9	9	9	9	5	
n-Nonane	9	9	9	9	9	9	9	16	16
n-Decane	9	9	9	9	9	9	9	16	16
n-Undecane	9	9	9	9	6	9	9	16	16
n-Dodecane	9	9	9	9	6	9	9	16	16
n-Tridecane			3	11		11	3	16	16
n-Tetradecane	6	6		11	6	6	3	16	16
n-Pentadecane	3		3	11		11	3	16	16
n-Hexadecane	6	6	3	11	6	6	3	16	16
n-Heptadecane			3	11		11	3	16	16
n-Octadecane	6	6	3	11	6	6	3	16	16
n-Nonadecane	3		3	11		11	3	16	16
n-Eicosane	6	6	3	11	6	11	3	16	16
n-Heneicosane									
n-Hexacosane	6	6			6				

Compound	Molecular Weight	S mg/L	Specific Gravity	VP atm	H(c/c)	log K_{ow}	BP C	Dair cm²/s	Dwater cm²/s
Branched Chain Alkanes									
Isobutane	9	9	9	9	9	9	9	5	
2,2-Dimethylbutane	9	9	9	9	9	9	9	16	16
2,3-Dimethylbutane	9	9	9	9	9	9	9	16	16
2,2,3-Trimethylbutane	9	9	9	9	9	11	9	16	16
2,2,3,3-Tetramethylbutane	9		9	9		11	9	16	16
Neopentane	9	9	9	9	9	9	9	16	16
Isopentane	9	9	9	9	9	11	9	5	
2-Methylpentane	9	9	9	9	9	11	9	5	
3-Methylpentane	9	9	9	9	9	9	9	16	16
3-Ethylpentane						11		16	16
2,2-Dimethylpentane	9	9	9	9	9	11	9	16	16
2,4-Dimethylpentane	9	9	9	9	9	11	9	16	16
2,3-Dimethylpentane	9	9	9	9	9	11	9	16	16
3,3-Dimethylpentane	9	9	9	9	9	11	9	16	16
2,2,3-Trimethylpentane						11		16	16
2,2,4-Trimethylpentane	9	9	9	9	9	11	9	16	16
2,3,3-Trimethylpentane						11		16	16
2,3,4-Trimethylpentane	9	9	9	9	9	11	9	16	16
2-Methyl-3-ethylpentane						11			16
2,4-Dimethyl-3-ethylpentane						11			
2-Methylhexane	9	9	9	9	9	12	9	12	12

Compound	Molecular Weight	S mg/L	Specific Gravity	VP atm	H(c/c)	log K_{ow}	BP C	Dair cm²/s	Dwater cm²/s
3-Methylhexane	9	9	9	9	9	11	9	16	16
2,2-Dimethylhexane						11		16	16
2,3-Dimethylhexane	4	3	3			11	4	16	16
2,4-Dimethylhexane						11		5	
2,5-Dimethylhexane						11		16	16
3,3-Dimethylhexane						11		16	16
3,4-Dimethylhexane						11		16	16
3-Ethylhexane						11		16	16
2-Methyl-3-ethylhexane						11			
2,2,4-Trimethylhexane						11		5	
2,2,5-Trimethylhexane	9	9	9	9	9	11	9	16	
2,3,3-Trimethylhexane						11			
2,3,5-Trimethylhexane						11			
2,4,4-Trimethylhexane						11			
2-Methylheptane	9	9	9	9	9	11	9	16	16
3-Methylheptane	6	6	9	9	9	11	9	16	16
4-Methylheptane						11		16	16
2,2-Dimethylheptane	3		3			11	3		
2,3-Dimethylheptane						11			
2,5-Dimethylheptane	4		4			11	4		
2,6-Dimethylheptane						11			
3,3-Dimethylheptane						11			

Compound	Molecular Weight	S mg/L	Specific Gravity	VP atm	H(c/c)	log K_{ow}	BP C	Dair cm^2/s	Dwater cm^2/s
3,4-Dimethylheptane						11			
2,2,4-Trimethylheptane						11			
2,4,6-Trimethylheptane						11			
3,3,5-Trimethylheptane						11			
3-Ethylheptane						11			
4-Ethylheptane						11			
2-Methyloctane						11			
3-Methyloctane	9	9	9	9		11	9		
4-Methyloctane	6	6	9	9	6	11	9		
2,6-Dimethyloctane						11			
2-Methylnonane						11			
3-Methylnonane						11			
4-Methylnonane						11			
4-Methyldecane									
2-Methyldecane									
2,6-Dimethyldecane									
2-Methylundecane									
2,6-Dimethylundecane									
Cycloalkanes									
Cyclopentane	9	9	9	9	9	9	9	16	16
Methylcyclopentane	9	9	9	9	9	9	9	16	16
1-Methyl-cis-2-ethylcyclopentane						11			
1-Methyl-trans-3-ethylcyclopentane						11			

Compound	Molecular Weight	S mg/L	Specific Gravity	VP atm	H(c/c)	log K_{ow}	BP C	Dair cm²/s	Dwater cm²/s
1,1-Dimethylcyclopentane						11			
1-cis-2-Dimethylcyclopentane						11			
1-trans-2-Dimethylcyclopentane						11			
1-cis-3-Dimethylcyclopentane						11			
1-trans-3-Dimethylcyclopentane						11			
1,1,2-Trimethylcyclopentane						11			
1,1,3-Trimethylcyclopentane	9	9	9	9	9	11	9		
1-trans-2-cis-3-Trimethylcyclopentane						11			
1-trans-2-cis-4-Trimethylcyclopentane						11			
1-trans-2-trans-4-Trimethylcyclopentane			10	10	10	11	10		
Ethylcyclopentane	9		9	9		11	9	16	
n-Propylcyclopentane	9	9	9	9	9	11	9	16	16
Isopropylcyclopentane						11			
1-cis-3-Dimethylcyclohexane						11			
1-trans-2-Dimethylcyclohexane	9	9	9	9	9	11	9		
1-trans-3-Dimethylcyclohexane						11			16
1-trans-4-Dimethylcyclohexane	9	6	6	9	9	11	9		
Ethylcyclohexane	9		9			11	9		
Cyclohexane	9	9	9	9	9	9	9	12	12
Methylcyclohexane	9	9	9	9	9	9	9	5	16
1-Methyl-2-ethylcyclohexane						11			
1-Methyl-3-ethylcyclohexane						11			

Compound	Molecular Weight	S mg/L	Specific Gravity	VP atm	H(c/c)	log K_{ow}	BP C	Dair cm^2/s	Dwater cm^2/s
1-Methyl-4-ethylcyclohexane						11			
1,3,5-Trimethylcyclohexane						11			
1,1,3-Trimethylcyclohexane	9	9	9	9	9	11	9		16
n-Butylcyclohexane						11			
n-Propylcyclohexane						11			
Hexylcyclohexane						11			
Heptylcyclohexane						11			
Pentylcyclopentane	9	9	9	9	9	11			
Straight Chain Alkenes									
Propylene						11			
cis-2-Butene						11			
trans-2-Butene						11			
1-Pentene	9	9	9	9	9	9	9	5	16
1-Pentyne	9	9	9	9	9	9	9	16	16
cis-2-Pentene					9	11			
trans-2-Pentene						11			
1-Hexene	9	9	9	9	9	9	9	5	16
1-Hexyne	9	9	9	9	9	9	9		
cis-2-Hexene						11			
trans-2-Hexene						11			
cis-3-Hexene						11			
trans-3-Hexene						11			

Compound	Molecular Weight	S mg/L	Specific Gravity	VP atm	H(c/c)	log K_{ow}	BP C	Dair cm²/s	Dwater cm²/s
cis-3-Heptene						11			
trans-2-Heptene	9	9	9	9	9	6	9		
1-Octene	9	9	9	9	9	9	9	16	16
1-Nonene	9	9	9	9	9	9	9	16	16
1-Decene	9	9	9	9	9	9	9	16	16
Tridecene						11			
Branched Chain Alkenes									
2-Methyl-1-butene						11			
3-Methyl-1-butene	9	9	9	9	9	11	9	16	
2-Methyl-2-butene						11			
2,3-Dimethyl-1-butene						11			
2-Methyl-1-pentene	9	9	9	9	9	11	9		
2,3-Dimethyl-1-pentene						11			
2,4-Dimethyl-1-pentene						11			
4,4-Dimethyl-1-pentene						11			
2-Methyl-2-pentene						11			
3-Methyl-cis-2-pentene						11			
3-Methyl-trans-2-pentene						11			
4-Methyl-cis-2-pentene						11			
4-Methyl-trans-2-pentene						11			
4,4-Dimethyl-cis-2-pentene						11			
4,4-Dimethyl-trans-2-pentene						11			
3-Ethyl-2-pentene						11			

Compound	Molecular Weight	S mg/L	Specific Gravity	VP atm	H(c/c)	log K_{ow}	BP C	Dair cm^2/s	Dwater cm^2/s
Cycloalkenes									
Cyclopentene	9	9	9	9	9	11	9		
3-Methylcyclopentene						11			
Cyclohexene	9	9	9	9	9	9	9	16	
Alkyl Benzenes									
Benzene	7	7	7	7	7	7	7	12	12
Toluene	7	7	7	7	7	7	7	12	12
Ethylbenzene	7	7	7	7	7	7	7	12	12
o-Xylene	7	7	7	7	7	7	7	12	12
m-Xylene	7	7	7	7	7	7	7	12	12
p-Xylene	7	7	7	7	7	7	7		
Styrene	9	9	9	9	9	9	9	12	12
1-Methyl-4-ethylbenzene	7	7	7	7	7	7	7	13	
1-Methyl-2-ethylbenzene	7	7	7	7	7	7	7	16	
1-Methyl-3-ethylbenzene	7		7	7		11	7	16	
1-Methyl-2-n-propylbenzene						11			
1-Methyl-3-n-propylbenzene						11			
1-Methyl-2-isopropylbenzene						11			
1-Methyl-3-isopropylbenzene						11			
1-Methyl-4-isopropylbenzene	7	7	7	7	7	7	7		
1-Methyl-3-t-butylbenzene						11			

Compound	Molecular Weight	S mg/L	Specific Gravity	VP atm	H(c/c)	log K_{ow}	BP C	Dair cm²/s	Dwater cm²/s
1-Methyl-4-t-butylbenzene						11			
1,2-Dimethyl-3-ethylbenzene						11			
1,2-Dimethyl-4-ethylbenzene						11			
1,3-Dimethyl-2-ethylbenzene						11			
1,3-Dimethyl-4-ethylbenzene						11			
1,3-Dimethyl-5-ethylbenzene									
1,4-Dimethyl-2-ethylbenzene						11			
1,3-Dimethyl-5-t-butylbenzene						11			
1,2,3-Trimethylbenzene	7	7	7	7	7	7	7	16	
1,2,4-Trimethylbenzene	7	7	7	7	7	7	7	16	
1,3,5-Trimethylbenzene	7	7	7	7	7	7	7	5	
1,2,3,4-Tetramethylbenzene	7		7	7		7	7		
1,2,3,5-Tetramethylbenzene	7		7	7		7	7		
1,2,4,5-Tetramethylbenzene	7	7	7	7	7	7	7	16	
1,2-Diethylbenzene						11			
1,3-Diethylbenzene						11			
1,4-Diethylbenzene						11		5	
1,2,4-Triethylbenzene						11			
1,3,5-Triethylbenzene						11			
n-Propylbenzene	7	7	7	7	7	7	7	16	
Isopropylbenzene	7	7	7	7	7	7	7	12	12
n-Butylbenzene	7	7	7	7	7	7	7	16	

Compound	Molecular Weight	S mg/L	Specific Gravity	VP atm	H(c/c)	log K_{ow}	BP C	Dair cm^2/s	Dwater cm^2/s
Isobutylbenzene	7	7	7	7	7	7	7	16	
sec-Butylbenzene	7	7	7	7	7		7		
t-Butylbenzene	7	7	7	7		7	7		
1-t-Butyl-3,4,5-trimethylbenzene						11			
n-Pentylbenzene	7	7	7	7	7	7	7		
Isopentylbenzene						11			
n-Hexylbenzene	7	7	7	7	7	7	7		
n-Heptylbenzene						11			
n-Octylbenzene						11			
4-Methylbiphenyl	8	8	8			8	8		
4,4'-Dimethylbiphenyl	8	8	8			8	8		
Phenylcyclohexane						11			
Naphtheno Benzenes									
Acenaphthene	8	8	8	8	8	8	8	19	19
Acenaphthylene	8	8	8	8	8	8	8	19	19
Indan	8	8		8	8	8	8		
1-Methylindan						11			
2-Methylindan						11			
4-Methylindan						11			
5-Methylindan						11			
Tetralin (tetrahydronaphthalene)	9	9	9	9	9	9	9		
5-Methylthtrohydronaphthalene						11			

Compound	Molecular Weight	S mg/L	Specific Gravity	VP atm	H(c/c)	log K_{ow}	BP C	Dair cm²/s	Dwater cm²/s
6-Methyltetrahydronaphthalene						11			
Fluorene	8	8	8	8	8	8	8	19	19
1-Methylfluorene	8	8				8	8		
Fluoranthene	8	8	8	8	8	8	8	19	19
2,3- Benzofluorene	6	6				8	8		
1,2- Benzofluorene	6	6				8	8		
Benz(b)fluoranthene	8	8		8		8	8		
Benz(k)fluoranthene	8	8		8	8	8	8	19	19
Indeno (1,2,3-cd) pyrene	6	6			19	6		19	19
Alkyl Naphthalenes									
Naphthalene	8	8	8	8	8	8	8	12	12
1-Methylnaphthalene	8	8	8	8	8	8	8	16	
2-Methylnaphthalene	8	8	8	8	8	8	8	16	
1,3-Dimethylnaphthalene	8	8	8			8	8		
1,4-Dimethylnaphthalene	8	8	8	8	8	8	8		
1,5-Dimethylnaphthalene	8	8				8	8		
2,3-Dimethylnaphthalene	8	8	8	8	8	8	8		
2,6-Dimethylnaphthalene	8	8	8	8	8	8	8		
1-Ethylnaphthalene	8	8	8	8	8	8	8		
2-Ethylnaphthalene	8	8	8	8	8	11	8		
1,4,5-Trimethylnaphthalene	6	8		8	8	8	8		
1-Phenylnaphthalene						11			

Compound	Molecular Weight	S mg/L	Specific Gravity	VP atm	H(c/c)	log K_{ow}	BP C	Dair cm²/s	Dwater cm²/s
Polynuclear Aromatics									
Anthracene	8	8	8	8	8	8	8	19	19
2-Methyl anthracene	8	6	8			6	8		
9-Methyl anthracene	8	8			6	8	8		
2-Ethyl anthracene	6	8				8			
9,10-Dimethyl anthracene	8	8				8			
Phenanthrene	8	8	8	8	8	8	8	19	19
1-Methylphenanthrene	8	8				8	8		
Pyrene	8	8	8	8	8	8	8	19	19
1-Methylpyrene						11			
Benz(a)anthracene	8	8	8	8	8	8	8	12	12
Chrysene	8	6		8	6	6	8	19	19
Triphenylene	8	8	8	8	8	8	8		
5-Methylchrysene	6	6				6			
Benz(a)pyrene	8	8		8	8	8	8	12	12
Benz(e)pyrene	8	8		8	8	6	8		
Perylene	8	8	8		8	8	8		
3-Methylcholanthrene	8	8	8	8	8	8	8		
Benz(ghi)perylene	8	8		8	8	8	8	19	19
Picene	6	6				6			
1,2,5,6-Dibenz anthracene	6	6		8	6	6	8	19	19
Coronene	8	8				8	8		

References for Appendix B

1. CRC (1981). *Handbook of Chemistry and Physics*. Physical Constants of Organic Compounds C65-C695. CRC PRESS, Inc., Boca Raton, FL.
2. Merck (1983). *The Merck Index of Chemicals, Drugs, and Biologicals* 1-1000. Merck & Co, Inc., Rahway, NJ.
3. Verschueren, K. (1983). *Handbook of Environmental Data on Organic Chemicals,* pp. 137-1207. Van Nostrand Reinhold Company, Inc. New York.
4. *Dangerous Properties of Industrial Materials*. (1984). General Chemicals 52-3087. Van Nostrand Reinhold Co. New York.
5. Lyman, W.J., P.J. Reidy, and B. Levy (1992). "Contaminants Dissolved in Groundwater," *in Mobility and Degradation of Organic Contaminants in Subsurface Environments* pp.207-235. C.K. Smoley, Inc., Chelsea, MI.
6. Arthur D. Little, Inc. (1994). "Transport and Fate of Non-BTEX Petroleum Chemicals in Soil and Groundwater," 1.1-9.1. API Publication, Cambridge, MA.
7. Mackay, D., W. Y. Shui, and K.C. Ma (1992). *Illustrated Handbook of Physical-Chemical Properties and Environmental Fate for Organic Chemicals Vol. I.* Lewis Publishers, Chelsea, MI.
8. Mackay, D., W. Y. Shui, and K.C. Ma (1992). *Illustrated Handbook of Physical-Chemical Properties and Environmental Fate for Organic Chemicals Vol. II.* Lewis Publishers, Chelsea, MI.
9. Mackay, D., W. Y. Shui, and K.C. Ma (1993). *Illustrated Handbook of Physical-Chemical Properties and Environmental Fate for Organic Chemicals Vol. III.* Lewis Publishers, Chelsea, MI.
10. Montgomery, J.H. (1991). *Groundwater Chemicals Field Guide*. Lewis Publishers, Chelsea, MI.
11. Lyman, W.J., W.F. Rheel, and D.H. Rosenblatt (1990). *Handbook of Chemical Property Estimation Methods: Environmental Behavior of Organic Compounds.* McGraw Hill Book Company, New York.
12. U.S. EPA (1993). "Supplement F to Compilation of Air Pollutant Emission Factors. Volume I: Stationary Point and Area Sources." U.S. Environmental Protection Agency, Office of Air Quality Planning Standards, Washington, DC, July 1993.
13. Reid, R.C., J.M. Prausnitz, and T.K. Sherwood. (1977). *Properties of Gases and_Liquids, 3rd ed.,* McGraw-Hill, New York.
14. Olsen, R.L. and A. Davis (1990). "Predicting the fate and transport of organic compounds in groundwater, Part 1," *Hazardous Materials Control* 3(3):38-63.
15. LUFT (1988). *Leaking Underground Fuel Tank Manual: Guidelines for Site Assessment, Cleanup, and Underground Storage Tank Closure*. State of California Leaking Underground Fuel Tank Force, May 1988.
16. McCabe, W.L., J.C. Smith, and P. Harriot (1985). *Unit Operations of Chemical Engineering. 4th ed.,* McGraw-Hill, New York.
17. Smith, J.M. and H.C. Van Ness (1975). *Introduction to Chemical Engineering Thermodynamics. 3rd ed.,* McGraw-Hill, New York, NY.

18. Hansch, C. and A. Leo (1991). ClogP Program for Estimation of Octanol-Water Partition Coefficients, DAYLIGHT Chemical Information Systems, Inc., Irvine, CA.

19. Risk-Based Corrective Action Computer Based Tier 1 Worksheets and Spreadsheets. Version Feb. 1995. Reference 4 - U.S. EPA, 1989: Hazardous Waste Treatment, Storage, and Disposal Facilities-U.S. EPA OAQPS, Air Emission Models (EPA-450/3-87-026).

20. ASTM (1995). *Standard Guide for Risk-Based Corrective Action Applied at Petroleum Release Sites.* Designation: E 1739-95. American Society for Testing and Materials, West Conshohocken, PA.